T0315102

Nuclear Electric Power

Nuclear Electric Power

Safety, Operation, and Control Aspects

J. Brian Knowles

Cover Design: Wiley
Cover Photography: © sleepyfellow/Alamy

Copyright © 2014 by John Wiley & Sons, Inc. All rights reserved.

Published by John Wiley & Sons, Inc., Hoboken, New Jersey.
Published simultaneously in Canada.

No part of this publication may be reproduced, stored in a retrieval system, or transmitted in any form or by any means, electronic, mechanical, photocopying, recording, scanning, or otherwise, except as permitted under Section 107 or 108 of the 1976 United States Copyright Act, without either the prior written permission of the Publisher, or authorization through payment of the appropriate per-copy fee to the Copyright Clearance Center, Inc., 222 Rosewood Drive, Danvers, MA 01923, (978) 750-8400, fax (978) 750-4470, or on the web at www.copyright.com. Requests to the Publisher for permission should be addressed to the Permissions Department, John Wiley & Sons, Inc., 111 River Street, Hoboken, NJ 07030, (201) 748-6011, fax (201) 748-6008, or online at http://www.wiley.com/go/permission.

Limit of Liability/Disclaimer of Warranty: While the publisher and author have used their best efforts in preparing this book, they make no representations or warranties with respect to the accuracy or completeness of the contents of this book and specifically disclaim any implied warranties of merchantability or fitness for a particular purpose. No warranty may be created or extended by sales representatives or written sales materials. The advice and strategies contained herein may not be suitable for your situation. You should consult with a professional where appropriate. Neither the publisher nor author shall be liable for any loss of profit or any other commercial damages, including but not limited to special, incidental, consequential, or other damages.

For general information on our other products and services or for technical support, please contact our Customer Care Department within the United States at (800) 762-2974, outside the United States at (317) 572-3993 or fax (317) 572-4002.

Wiley also publishes its books in a variety of electronic formats. Some content that appears in print may not be available in electronic formats. For more information about Wiley products, visit our web site at www.wiley.com.

***Library of Congress Cataloging-in-Publication Data*:**

Knowles, J. B. (James Brian), 1936-
Nuclear electric power : safety, operation and control aspects/J.B. Knowles.
pages cm
"Published simultaneously in Canada"–Title page verso.
Includes bibliographical references and index.
ISBN 978-1-118-55170-7 (cloth)
1. Nuclear power plants. 2. Nuclear reactors–Safety measures. 3. Nuclear reactors–Control. 4. Nuclear energy. 5. Electric power systems. I. Title.
TK1078.K59 2013
621.48'3–dc23

2013000147

10 9 8 7 6 5 4 3 2 1

To Lesley Martin
A good neighbor to everyone and our dear friend.

Contents

Preface

If the industries and lifestyles of economically developed nations are to be preserved, then their aging, high-capacity power stations will soon need replacing. Those industrialized nations with intentions to lower their carbon emissions are proposing nuclear and renewable energy sources to fill the gap. As well as UK nuclear plant proposals, China plans an impressive 40% new-build capacity, with India, Brazil, and South Korea also having construction policies. Even with centuries of coal and shale-gas reserves, the United States has recently granted a construction license for a pressurized water reactor (PWR) near Augusta, Georgia. Nuclear power is again on the global agenda.

Initially renewable sources, especially wind, were greeted with enthusiastic public support because of their perceived potential to decelerate global climate change. Now however, the media and an often vociferous public are challenging the green credentials of all renewables as well as their ability to provide reliable electricity supplies. Experienced engineering assessments are first given herein for the commercial use of geothermal, hydro, solar, tidal and wind power sources in terms of costs per installed MW, capacity factors, hectares per installed MW and their other environmental impacts. These factors, and a frequent lack of compatibility with national power demands, militate against these power sources making reliable major contributions in some well-developed economies. Though recent global discoveries of significant shale and conventional gas deposits suggest prolonging the UK investment in reliable and high thermal efficiency combined cycle gas turbine (CCGT) plants, ratified emission targets would be contravened and there are also political uncertainties. Accordingly, a nuclear component is argued as necessary in the UK Grid system. Reactor physics, reliability and civil engineering costs reveal that water reactors are the most cost-effective. By virtue of higher linear fuel ratings and the emergency cooling option provided by separate steam generators, PWRs are globally more widely favored.

Power station and grid operations require the control of a number of system variables, but this cannot be engineered directly from their full nonlinear dynamics. A linearization technique is briefly described and then applied to successfully establish the stability of reactor power, steam drum-water level, flow in boiling reactor channels and of a Grid network as a whole. The reduction of these multivariable problems to single input-single output (SISO) analyses illustrates the importance of specific engineering insight, which is further confirmed by the subsequently presented nonlinear control strategy for a station blackout accident.

Public apprehensions over nuclear power arise from a perceived concomitant production of weapons material, the long-term storage of waste and its operational safety. Reactor physics and economics are shown herein to completely separate the activities of nuclear power and weapons. Because fission products from a natural fission reactor some 1800 million years ago are still incarcerated in local igneous rock strata, the additional barriers now proposed appear more than sufficient for safe and secure long-term storage. Spokespersons for various non-nuclear organizations frequently seek to reassure us with "Lessons have been learned": yet the same misadventures still reoccur. Readers find here that the global nuclear industry has indeed learned and reacted constructively to the Three Mile Island and Chernobyl incidents with the provision of safety enhancements and operational legislation. With regard to legislation, the number of cancers induced by highly unlikely releases of fission products over a nuclear plant's lifetime must be demonstrably less than the natural incidence by orders of magnitude. Also the most exposed person must not be exposed to an unreasonable radiological hazard. Furthermore, a prerequisite for operation is a hierarchical management structure based on professional expertise, plant experience and mandatory simulator training. Finally, a well-conceived local evacuation plan must pre-exist and the aggregate probability of all fuel-melting incidents must be typically less than 1 in 10 million operating years.

Faulty plant siting is argued as the reason for fuel melting at Fukushima and not the nuclear technology itself. If these reactors like others had been built on the sheltered West Coast, their emergency power supplies would not have been swamped by the tsunami and safe neutronic shut-downs after the Richter-scale 9 quake would have been sustained.

To quantify the expectation of thyroid cancers from fission product releases, international research following TMI-2 switched from intact plant performance to the phenomenology and consequences of fuel melting (i.e., Severe Accidents) after the unlikely failure of the multiple emergency core cooling systems. This book examines in detail the physics, likelihood and plant consequences of thermally driven explosive interactions between molten core debris and reactor coolant (MFCIs). Because such events or disintegrating plant items, or an aircraft crash are potential threats to a reactor vessel and its containment building, the described "replica scale" experiments and finite element calculations were undertaken at Winfrith. Finally, the operation and simulation of containment sprays in preventing an over-pressurization are outlined in relation to the TOSQAN experiments.

This book has been written with two objectives in mind. The first is to show that the safety of nuclear power plants has been thoroughly researched, so that the computed numbers of induced cancers from plant operations are indeed orders of magnitude less than the natural statistical incidence, and still far less than deaths from road traffic accidents or tobacco smoking. With secure waste storage also assured, voiced opposition to nuclear power on health grounds appears irrational. After 1993 the manpower in the UK nuclear industry contracted markedly leaving a younger minority to focus on decommissioning and waste classification. The presented information with other material was then placed in the United Kingdom Atomic Energy Authority (UKAEA) archives so it is now difficult to access. Accordingly this compilation under one cover is the second objective. Its value as part of a comprehensive series of texts remains as strong as when originally conceived by the UKAEA. Specifically, an appreciation helps foster a productive interface between diversely educated new entrants and their experienced in situ industrial colleagues.

Though the author contributed to the original research work herein, it was only as a member of various international teams. This friendly collaboration with UKAEA, French, German and Russian colleagues greatly enriched his life with humor and scientific understanding. Gratitude is also extended to the Nuclear Decommissioning Authority of the United Kingdom for their permission to reproduce, within this book alone, copyrighted UKAEA research material. In addition thanks are due to Alan Neilson, Paula Miller, and Professor Derek Wilson, who have particularly helped to "hatch" this book. Finally, please note that the opinions expressed are the author's own which might not concur with those of the now-disbanded UKAEA or its successors in title.

Brian Knowles

River House, Caters Place, Dorchester

Glossary

AEC	Atomic Energy Commission (US)
AEEW	Atomic Energy Establishment Winfrith
AERE	Atomic Energy Research Establishment (Harwell)
AGR	Advanced Gas Cooled Reactor
ALARP	As Low as Reasonably Practicable
ANL	Argonne National Laboratory (US)
ASME	American Society of Mechanical Engineers
AWRE	Atomic Weapons Research Establishment (Aldermaston)
BNES	British Nuclear Energy Society
BRL	Ballistics Research Laboratory (US)
BWR	Boiling Water Reactor
CEGB	Central Electricity Generating Board (now disbanded)
CEN	Centre d'Etude Nucléaires (Grenoble)
CFR (EFR)	Proposed Commercial (European) Fast Reactor
Corium	A mixture of fuel, clad and steelwork formed after core-melting in a Severe Accident
DBA	Design Base Accident(s)
EC	European Commission
ECCS	Emergency Core-Cooling Systems
EWEF	Each Way-Each Face (for steel reinforcement of concrete)
HCDA	Hypothetical Core Disruptive Accident (⇔ Severe Accident)
HMSO	Her Majesty's Stationary Office (London)
IAEA	International Atomic Energy Agency
IEE	Institute of Electrical Engineers (now IET)
IEEE	Institute of Electrical and Electronic Engineers (US)
JRC	(European) Joint Research Centre (Ispra)
KfA	Kernforschungsanlage (Jülich)

KfK	Kernforschungszentrum Karlsruhe (now Institut für Neutronenphysik)
LMFBR	Liquid Metal Fast Breeder Reactor
L(S)LOCA	Large (Small) Loss of Coolant Accident
MCR	Maximum Continuous Rating or Installed Capacity (MW or GW)
MFCI	Molten Fuel Coolant Interaction
MFTF	Molten Fuel Test Facility (at AEEW)
MIMO	Multi Input-Multi Output (dynamic system)
NNC	National Nuclear Corporation (UK)
NRDC	National Research Defense Council (US)
NUREG	Nuclear Regulatory Commission (US)
OECD	Organization for Economic Cooperation and Development
ORNL	Oak Ridge National Laboratory (US)
PFR	Prototype Fast Reactor (UK)
PWR	Pressurized Water Reactor
SISO	Single Input-Single Output (dynamic system)
SGHWR	Steam Generating Heavy Water Reactor (at AEEW)
SNUPPS	Standard Nuclear Unit Power Plant System (Westinghouse US)
STP	Standard Temperature and Pressure
TCV	Turbine Control Valve (steam)
UMIST	University of Manchester Institute of Science and Technology
UKAEA	United Kingdom Atomic Energy Authority

Principal Nomenclature

η	An efficiency
P	Power, pressure
$P_A; P(A)$	Probability of an event A
$P(B/A); P_{B/A}$	Conditional probability of B given that A has occurred
ω	Angular frequency
ρ	Density; core reactivity
T	Temperature
T^*	Reactor period
$\sum$	Macroscopic cross-section; an algebraic sum
s	Complex variable of the Laplace transformation
$x(t)$	The state vector of a finite number of Laplace transformable functions
$\dot{x}$	Total temporal derivative of x
$\bar{x}$	Upper bar denotes the Laplace transform of $x(t)$
$\{A, B, C, D\}$	State Space matrices
D	Hydraulic diameter; a characteristic length; radiological dose
det	determinant of
A^{-1} or $\hat{A}$	Inverse of a matrix A
I	Identity matrix; specific internal energy
λ	Eigenvalue; neutronic lifetime; a wavelength
i	$= \sqrt{-1}$
Re	Real part of a complex number; Reynolds number
j, k, m, n	Non-negative integers
$C_p; (C_v)$	Specific heat at constant pressure; (volume)
γ	s-plane contour capturing all unstable poles; or C_p/C_v
ϕ	Angular phase difference; neutron flux; heat flux
∇	Vector differential operator

δ	Prefixing an infinitesimal change in a variable
Δ	Prefixing a sizeable change in a variable
W	Mass flow rate; a mass creation rate (e.g., of fragments); wind factor
G	Mass flux = mass flow per unit area
ν	Specific volume $= 1/\rho$
ϵ	Thermal emissivity; induced mechanical strain
σ	Stefan-Boltzmann constant; condensation coefficient; Statistical standard deviation; volumetric heat generation rate
α	Thermal diffusivity
κ	Thermal conductivity
V	Velocity
$\mathcal{G}$	Gruneisen function
$erfc$	Complementary error function
Z	Acoustic impedance
h	A heat transfer coefficient
E	Energy
$\mathcal{E}$	Statistical expectation of the associated variable
$U(t)$	Unit step function $= 1$ for $t > 0$ but 0 otherwise
g	Gravitational acceleration
μ	Dynamic viscosity
Nu; Pr	Nusselt; Prandtl number
$\in$	Belonging to
$\mathbb{R}$	Set of all real numbers
$\triangleq$	Equality by definition: not deducible

The diverse range of subjects with the preferred use of conventional symbolism makes multiple connotations inevitable, but local definitions prevent ambiguity. All vector variables are embolded.

CHAPTER 1

Energy Sources, Grid Compatibility, Economics, and the Environment

1.1 BACKGROUND

If the industries and accustomed lifestyles of the economically well-developed nations are to be preserved, their aging high-capacity ($\gtrsim 100\,\text{MW}$) electric power plants will soon require replacement with reliable units having lower carbon emissions and environmental impacts. Legally binding national targets [1] on carbon emissions were set out by the European Union in 2008 to mitigate their now unequivocal effect on global climate change. In 2009, the UK's Department of Energy and Climate Change [1] announced ambitious plans for a 34% reduction in carbon emissions by 2020. The principal renewable energy sources of Geothermal, Hydro-, Solar, Tidal and Wind are now being investigated worldwide with regard to their contribution towards a "greener planet." Their economics and those for conventional electricity generation are usually compared in terms of a Levelized Cost which is the sum of those for capital investment, operation, maintenance and decommissioning using Net Present-day Values. Because some proposed systems are less well-developed for commercial application (i.e., riskier) than others, or are long term in the

Nuclear Electric Power: Safety, Operation, and Control Aspects, First Edition.
J. Brian Knowles.
© 2014 John Wiley & Sons, Inc. Published 2014 by John Wiley & Sons, Inc.

sense of capitally intensive before any income accrues, the now necessary investment of private equity demands a matching cash return [52]. Also in this respect the electric power output from any generator has a degree of intermittency measured by

$$\text{Capacity Factor} \triangleq (\text{Annual Energy Output})/(\text{Annual Output at Max. Power}) \tag{1.1}$$

These aspects are included as discounted cash flows in a Capital Asset Pricing Model that assesses the commercial viability of a project with respect to its capital repayment period.

As well as satisfactory economics and environmental impact, a replacement commercial generator in a Grid system must provide its centrally scheduled contribution to the variable but largely predictable power demands on the network. Figure 1.1 illustrates such variable diurnal and seasonal demands in the United Kingdom. It is often

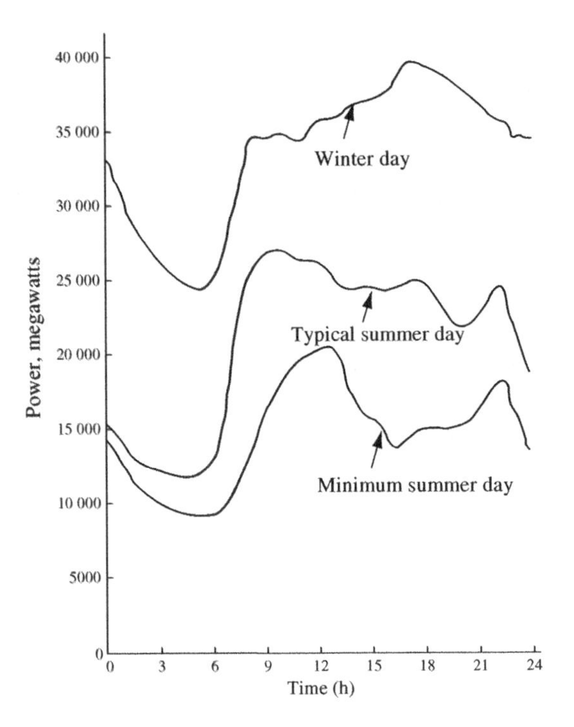

Figure 1.1 Typical Electrical Power Demands in the United Kingdom

claimed in the popular media that a particular wind or solar installation can provide a specific fraction of the UK's electrical energy demand (GWh), or service so many households. Often these energy statistics are based on unachievable continuous operation at maximum output and an inadequate instantaneous power of around 1½ kW per household.[1] As explained in Section 3.3 it is crucial to maintain a close match between instantaneous power generated and that consumed: as otherwise area blackouts are inevitable. Moreover, because these renewables fail to deliver their quotas under not improbable weather conditions, additional capital expenditure is necessary in the form of reliable backup stations. Assessments of the economics, reliability, Grid compatibility and environmental impacts of commercially sized generating sources now follow.

1.2 GEOTHERMAL ENERGY

Geothermal energy stems from impacts that occurred during the accretive formation of our planet, the radioactive decay of its constituents and incident sunlight. Its radioactive component is estimated [2] as about 30 TW, which is about half the total and twice the present global electricity demand. However, commercial access is achievable only at relatively few locations along the boundaries of tectonic plates and where the geology is porous or fractured. Though hot springs and geysers occur naturally, commercial extraction for district heating, horticulture or electric power involves deep drilling into bedrock with one hole to extract hot water and another thermally distant to inject its necessary replenishment. There are presently no commercial geothermal generation sites in the United Kingdom, but a 4½ km deep 10 MW station near Truro is under active consideration.

The Second Law of Thermodynamics [3] by Lord Kelvin asserts that a heat engine must involve a heat source at a temperature T_1 and a cooler heat sink at a temperature T_0. In 1824, Carnot proved that the maximum efficiency η^* by which heat could be converted into mechanical work is

$$\eta^* = 1 - T_0/T_1 \quad \text{with} \quad T_1, T_0 \, \text{in Kelvin} \qquad (1.2)$$

[1] A typical electric kettle consumes 2 kW.

Given a relatively hot geothermal source of 200°C and a condensing temperature of 40°C, the above efficiency bound evaluates as 34%, but intrinsic thermodynamic irreversibilities [3] allow practical values [2] of only between 10 and 23%. Because the majority of geothermal sources have temperatures below 175°C they are economic only for district and industrial space heating or as tourist spectacles in areas of outstanding beauty (e.g., Yosemite National Park, USA). Exploitation of the higher temperature sources for electric power is engineered by means of a Binary Cycle system, in which extracted hot water vaporizes butane or pentane in a heat exchanger to drive a turbo-alternator. Replenishment water for the geothermal source is provided by the colder outlet, and district or industrial space heating is derived from recompression of the hydrocarbon. The largest geothermal electricity units are located in the United States and the Philippines with totals of 3 and 2 MW, respectively, but these countries with others intend further developments.

According to the US Department of Energy an 11 MW geothermal unit of the Pacific Gas and Electric Company had from 1960 an operational life of 30 years, which matches those for some fossil and nuclear power stations. Because geothermal generation involves drilling deep into bedrock with only a 25 to 80% chance of success, development is both risky and capital intensive and so it incurs a high discount rate. Moreover, despite zero fuel charges, low thermal conversion efficiencies reduce the rate of return on invested capital, which further increases interest rate repayments. That said, nations with substantial geothermal resources are less dependent on others for their electricity which is an important political and economic advantage. Construction costs for a recent 4.5 MW unit in Nevada, the United States were $3.2M per installed MW.

Geothermal water contains toxic salts of mercury, boron, arsenic and antimony. Their impact on a portable water supply is minimized by replenishments at similar depths to the take-off points. These sources deep inside the earth's crust also contain hydrogen sulfide, ammonia and methane, which contribute to acid rain and global warming. Otherwise with an equivalent carbon emission of just 122 kg per MWh, geothermal generation's "footprint" is small compared with fossil-fired production. However, the extraction process fractures rock strata that has caused subsidence around Wairakei, NZ, and at Basel CH small Richter-scale 3.4 earth tremors led to suspension of the project after just 6 days.

Geothermal energy for domestic and small-scale industrial space heating can be provided without an environmental impact by heat pumps [3,15]. An early 1920's example is the public swimming pool at Zürich CH which used the River Limmat as its heat source. Finally, some recently built UK homes have heat pumps whose input is accessed from coils buried in their gardens.

1.3 HYDROELECTRICITY

Some 715 GW of hydroelectric power are already installed worldwide, and in 2006, it supplied 20% of the global electricity demand and 88% of that from all renewable sources [4]. Large schemes of more than about 30 MW involve the construction of a convex dam across a deep river gorge whose sides and bottom must be geologically sound. In addition, a sufficiently large upstream area must exist for water storage (i.e., availability) and sufficient precipitation or glacial melt must be available to maintain this reservoir level. Viable large hydroelectric sites thus necessitate a special topography and geology, but are nevertheless more numerous and powerful than geothermal ones as indicated by Table 1.1. Both renewable sources, however, are reliable and can accommodate the variations in power demanded by an industrialized economy. Water below a dam is drawn-off in large pipes (penstocks) to

Table 1.1
Some Annual Energy Consumptions and Dams in 2006

Country	United Kingdom	United States	China	Brazil	Norway	Egypt
Energy pa (GWh)	0.345E6	3.87E6	3.65E6	0.403E6	0.110E6	0.849E6
% Hydro	1.3	9.9	17.0	25	99	~15
Dam (GW)	Pitlochry 0.245	Grand Coulee 6.8	Three Gorges 22.5	Itaipu[a] 14.0	Rjukan 0.06	Aswan 2.1
Completed	1951	1942	2010	1991	1911	1970

[a]Shared with Paraguay.

drive vertically mounted turbines whose blades are protected from cavitation by a slightly rising outfall to downstream [10].

Formal legislation on carbon emissions [1] and the increasing costs of fossil fuels have been driving global construction programs for hydro-electricity. Suitable large-scale sites in the United Kingdom were fully developed during 1940–1950, and future opportunities will focus on small or microscale plants (< 20 MW) whose total potential is estimated at 3% of national consumption [5]. Redundant factories from the UK's industrial revolution provide opportunities for microgeneration like the 50 kW rated plant at Settle [6], but even after a copious rainfall the claim to supply 50 homes is optimistic. It is to be concluded that no large-scale hydro-sources are available now to compensate materially for the impending demise of the UK's aging fossil and nuclear power stations. The situation [21] in the United States is that large and small-scale hydro-generation have remained largely unchanged over the past 10 years and that future renewable energy development will center on wind turbines [7].

Dams are sometimes breached by river spates or earthquakes despite the inclusion of such statistics in their design. For example environmental damage and a serious loss of life ensued from the failure of the Banqiao Dam [11] (China). Here there were 26,000 immediate fatalities and a further 145,000 from subsequent infections. No worse nuclear accident could be envisaged than that in 1986 of the RMBK reactor at Chernobyl which is designated 7 on the IAEA scale of 1 to 7. The 186 exposed settlements with a total population of some 116,000 were evacuated within 12–13 days. In the specific context of health issues, the International Chernobyl Project [13] of the IAEA reported

i. "Adverse health effects attributed to radiation have not been substantiated."
ii. "There were many psychological problems of related anxiety and stress."
iii. "No abnormalities in either thyroid stimulating hormone (TSH) or thyroid hormone (TH) were found in the children examined."

The earlier Three Mile Island accident (1979) did not directly cause any on or off-site fatalities, though some occurred from remote road accidents due to the absence of an organized evacuation plan. Historic catastrophic failures of large hydroelectric dams have thus caused far greater fatalities than the worst nuclear power plant accident, but their relative probabilities require of course quantification,[2] which must now account for the lessons learnt and practiced. Though all large dams are potential terrorist targets, the Ruhr-dam bombing raids in World War II demonstrate that success necessitates a scientifically sophisticated attack.

1.4 SOLAR ENERGY

Photoelectricity was discovered by Hallwachs [16] in 1888, and its quantum mechanical analysis was provided by Einstein in 1905. However, the necessary research toward viable electrical power units actually began in 1954 with transistor development by Bell System Laboratories NJ. Solar cells for this purpose are now [17] series-connected arrays of p–n junctions in ribbon polycrystalline silicon which have a quoted life expectancy of 30 years.[3] Though mono-crystalline devices offer a somewhat greater conversion efficiency of sunlight into electrical energy, ribbon technology is cheaper with a theoretical maximum conversion efficiency [17] of 29%. By manufacturing ever-thinner devices charge carrier recombination during diffusion has been reduced so as to achieve efficiencies of around 18%. Conversion losses also occur as a result of atmospheric or bird deposits and in the thyristor inverters between domestic and Grid networks. Because solar radiation has no cost, a low conversion efficiency principally aggravates capital investment and environmental impact.

During the four winter months Table 1.2 and Figure 1.1 show that the average of 1–2 sunshine hours around mid-day are well outside the UK's national peak demands between 1600 and 2100 h. Though solar cells provide some twilight output the 17% capacity factor for UK solar arrays from Table 1.2 suggests an inadequate annual return on capital for commercial plants. However, Spain and the United States lead the

[2] See Chapter 4 for the nuclear power plant case. For the Banqiao Dam, the probability of a storm created overflow was assessed [11] as 0.001 p.a., so it was considered safe for 1000 years.

[3] Experience indicates that semiconductors are most likely to fail in a short period after fabrication; hence a manufacturer's "burnin".

Table 1.2
Average Sunshine Hours Per Day [18] in the United Kingdom and Spain

Month	J	F	M	A	M	J	J	A	S	O	N	D
London	1	2	4	5	6	7	6	6	5	3	2	1
Madrid	5	6	6	8	9	11	12	11	9	6	5	5

world in the use of solar energy [19]. Spain has a currently installed capacity of 432 MW with plans for a total 900 MW, and the United States has presently 457 MW with a large 968 MW unit under construction in Riverside County, California. As well as Capacity Factors twice that of a UK plant, their solar output conveniently peaks with that of summer noon-time electricity demand for air conditioning. By avoiding the synchronization of low merit order[4] fossil-fired stations, Spanish and US solar units further enhance their economics and reduce carbon emissions. Moreover, these countries have large arid or otherwise unusable areas of land whose purchase offers no impediment to commercially sized developments. For example in the United States, a Boston plant [1] of 1.3 MW was recently built on the contaminated land of a derelict gas works with a utilization of 1.9 hactares per installed MW. On the other hand high land prices, climate and incompatibility with the national electricity demand further militate against solar generation in the United Kingdom. Indeed the United Kingdom reduced its subsidy for commercial generation [30] ($\geq$50 kW) in February 2011, and later in February 2012 attempted to cut the feed-in tariff for domestic roof-top units of a few kW.

1.5 TIDAL ENERGY

Tidal movements result from an interaction of the gravitational fields of the sun and moon with the earth's water masses. The global potential for tidal power is an estimated [22] 6 TW, but there are just a few special locations for its economic exploitation. Specifically they have the

[4] Fossil-fired stations rank in a merit order corresponding to their annually averaged generation costs per kWh.

appropriate orientation to access the Coriolis forces created by the earth's rotation and a shape whose natural oscillatory frequency closely matches that of the tides [23]. Depending on the situation, then principally[5] tidal range or tidal stream systems are the flexible means of extracting the energy. In the former mature technology, water flows at a flood tide are trapped behind a dam or lagoons and in the process drive horizontal low-head turbines during parts of both flood and ebb tides. Suitable sites include the Severn Estuary (UK), La Rance (France), Bay of Fundy (Canada) and some others where the tidal range exceeds the necessary 7 m for an economic development [22]. Though lagoons as a means of reducing upstream ecological damage were considered for the Severn Estuary scheme, they were subsequently rejected for the induced scouring of the interstitial seabed and their cost-effectiveness [23]. In fact there are no tidal lagoon schemes presently in the world.

Tidal stream devices extract a portion of the kinetic energy from relatively fast tidal currents as for example at the UK sites of Pentland Firth, Strangford Lough, and Alderney. Various blade designs are under development for the Kaplan-type turbines, but as yet no real choice exists with regard to efficiency and cost-effectiveness [23]. Individual units can provide up to 1 MW, so "farms" of as many as 30 are planned in order to justify the provision of substantial new Grid connections to prevent transmission "bottlenecks" that would otherwise exist between these generating units and the centers of largest electricity demand [23]. Other issues can be identified by assessing the potential of two powerful tidal steams in New Zealand to provide a material portion of its 13 GW demand [116]. These particular steams peak around 5 h apart, so their combined power profile assumes the form

$$P = P_0[2 + \cos\theta + \cos(\theta + \phi)] \tag{1.3}$$

where

$$\theta = \omega t; \quad \omega = \pi/6 \text{ and } \phi = 5\pi/6$$

[5] Wave devices so far appear unable to meet the sporadic violent storms in the United Kingdom and Australia.

Extreme values occur at

$$\tan\hat{\theta} = -b/a; \quad b = \sin\phi \quad \text{and} \quad a = 1 + b$$

giving the ratio

$$P_{\max} : P_{\min} = 1.33 : 1.0 \tag{1.4}$$

Granted sufficient tidal energies, equation (1.4) shows that even at minimum flow enough Kaplan turbines could meet a specific quota in any 24–h period. The excess power at higher flows could simply be rejected by de-exciting individual generators. Though apparently very promising, some 5000 of the presently largest 1 MW turbines would be required for 5 GW. Moreover, powerful tidal streams (~5 m/s) offer very restricted safe diving windows[6] and are likely to entrain amounts of grit that could erode turbine blades. Because installation, interconnections and maintenance do not benefit from scale only much smaller stations appear viable. In this respect the 30 MW trial at Strangford Lough should be definitive in terms of reliability, scale, and costs. In contrast the proposed 16-km-long Severn Barrage between Cardiff and Weston would have produced a material maximum output of 8.46 GW that is closer to the largest demand centers and 4.7% of UK national consumption.

Flood and ebb tides each occur twice daily and repeat about 1 h later each successive day. Also tidal ranges and stream speeds vary periodically over the 28-day lunar cycle due to the changing alignment of the sun and moon. The electrical output of a tidal generator varies in like manner so it is frequently out of step with a grid network's daily demand. Nevertheless tides are regular and largely predictable [24] so tidal generators can be readily incorporated into the largely predictable daily load schedule of a Grid network. Furthermore the relatively low height of a tidal barrage results in low mechanical stresses so that unforeseen outages are rare. Indeed the La Rance 240 MW plant has performed without a major incident for some 40 years since its construction in 1966. At 2006 prices, maintenance costs for the Severn scheme are estimated at £139M or $214M per year which is about 1% of total capital costs, and

[6] Two 40-minute periods per day around Alderney.

Table 1.3
Installation Costs for the Severn Barrage at 2006 Prices

	Civil Engineering	Turbines	Transmission
£M	9029	4198	2291
$M	13,890	6459	3523

at $1 = £0.65.

operational costs are minimal for the same reasons as with a conventional hydroelectric scheme. With zero "fuel" costs, the levelized cost for tidal barrage generation is essentially initial capital cost,[7] but the protracted construction period[8] without income attracts a high discount rate despite a lifetime expectancy of some 120 years. Itemized capital costs [23] for the proposed Severn Barrage are shown in Table 1.3. Capacity factors [22] accounting for the exploitable tidal ranges and high reliabilities lie globally in the range 20 to 35%. Accordingly, installation costs in Table 1.3 should be multiplied by 1/0.35 to give a minimum capital asset cost of £5.53M or $8.05M per MW which broadly indicates the necessary consumer charges for a viable return on the investment.[9] Tidal lagoons [23] for the Severn Estuary would each have had a maximum output of 60 MW with an installation cost of between £81.5M to £234M. However, on the same basis of a 35% capacity factor, their capital asset cost lies between $6M to $17M per MW. Using data in Ref. [23], corresponding estimates for tidal stream farms of 30 units with a 20-year life expectancy range from $5.3M to $12M per MW, but these technologies are as yet commercially unproven. In the above context, US nuclear power stations [25] achieved capacity factors of over 90% in 2007 to 2008. It appears unrealistic to contemplate the future decommissioning cost of a barrage a century later on, but their low height implies far less than for the high dams of conventional hydroelectric plant.

[7] Sometimes described as engineering procurement construction costs [26].

[8] Six years for La Rance and an estimated 12 years for the Severn [23].

[9] ENTEC argues that a viable tidal power investment necessitates installed or nameplate ratings of at least 2.8 GW.

Tidal barrages affect the physical, chemical and ecological features of an estuary. La Rance is the one materially sized plant in the world, but studies of its environmental impact have been limited [23]. However, insight can be gained from observations on harbor walls, jetties, bridges, or breakwaters. Though a barrage mitigates upstream flooding by tidal surges, its associated storage would regularly flood marshes used for previous centuries by wildlife, and it may also reflect tidal surges to damage downstream areas. Water management at La Rance creates the advertised "largest whirlpool in the world" and has become a frequently videoed tourist attraction. Also changes in hydrodynamics, dissolved oxygen and salt concentrations can producc radical changes in local flora and fauna as well as to sources of public drinking water. At La Rance for example sand eels and plaice have been replaced by sea bass and cuttlefish. Sediment accumulated behind a barrage not only requires regular dredging as in a conventional hydro scheme, but the estuary's topography and shipping channels could be altered by the modified silt deposits. A subdued water flow behind a barrage may also concentrate human and industrial effluents. The "footprint" of a tidal barrage scheme is highly significant for the densely populated United Kingdom, and that for La Rance is 9.38 hactares per installed MW.

Though La Rance was originally intended as a prototype of many tidal stations supplying a material portion of France's electricity requirements [27], nuclear power development became the final choice. Reasons for EDF's decision have apparently not been published [27], but the above environmental issues must have entered the argument. Whatever the truth, the nuclear choice became very profitable. After the unification of Germany in 1990 and with Chernobyl probably influencing matters, the Russian RMBK reactors and heavily polluting lignite-burning generators in the DDR were decommissioned. The resulting energy gap was filled by energy from France's largely[10] nuclear plants, which also buffer the UK's grid network by the 2 GW cross-channel ac connector. In the case of the Severn Barrage, BBC News announced on 9th February 2010 that the estuary would be "devastated" and on the 5th September the *Guardian* newspaper wrote that governmental support had been withdrawn.

[10] About 80% national capacity.

1.6 WIND ENERGY

Wind is a costless, inexhaustible but intermittent energy source. Early twentieth century wind turbines were domestic multibladed fixed-pitch metal machines. Though aviation developments during World War I led to higher efficiency two- and three-bladed rotors [28], output powers before 1945 were generally no greater than 5 KW. A notable exception was the 1.25 MW variable-pitch machine on Grandpa's Knob (Vermont, US), but after just 6 years in service one of its 8 ton metal blades fractured. Now lighter composite blades, power electronics and control schemes [29] result in mechanically reliable 2 to 3 MW rated machines with 20 to 30 year life expectancies. These are frequently deployed for individual factories or in "farms" of as many as 100 units to enable a cost-effective Grid network connection [30,31]. Powerful motivation for the present commercial developments of wind power emanates from the escalating costs of fossil fuels and the strident pandemic voices urging less global pollution.

An idealized fluid dynamic model [28] for the steady-state output power P of an isolated wind turbine reveals the engineering features necessary for materially sized power generation

$$P = \tfrac{1}{2} C_p \rho A V_1^3 \tag{1.5}$$

where

C_p – power coefficient; ρ – density of air, which at STP $=1.29\ \mathrm{kg/m^3}$

A – area swept by blades; V_1 – steady incident upstream windspeed

Because the density of air is so small[11] commercially sized power generation requires very large diameter blades. Contemporary 3 MW rated machines have 60 m diameter blades elevated to a total height of 115 m, and so are particularly visually intrusive [324]. Furthermore the 100 units of the Isle of Thanet 300 MW wind offshore wind farm [30,31] occupy 3500 hactares or about 3500 football pitches. Land shortages

[11] Output steam density for a PWR boiler is some two decades larger.

and therefore prices [20] in the United Kingdom dictate the construction of mainly offshore farms. In contrast, the complete nuclear reactor site at Winfrith occupied [32] about 3½ hactares and reliably[12] generated its rated 100 MW for around 25 years. Moreover, its structure was designed to be compact, and apart from a water vapor plume all was invisible[13] from the main A352 road about 1 km away.

Equation (1.5) also shows that accessible power is proportional to the cube of the incident wind speed, so the geographical location of a wind farm is very important. As early as 1948 a UK government committee organized a survey of national onshore windspeeds [28]. Since then many countries have produced their own contour maps of annual mean wind speeds (Isovents) with coastal and offshore regions appearing to be the most economically favorable. Though Grid connections using synchronous or induction generators [35] have not encountered insurmountable difficulties [28], UK offshore wind farms now generate three-phase rectified dc current with onshore Grid-tied inverters [33,34] so as to effect an efficient and more economic Grid connection. Specifically, cable costs are determined by both the peak transmitted voltage and current, as heating by the latter reduces the breakdown voltage of its insulation. For a given cable, the ac and dc powers transmitted are

$$P_{ac} = \tfrac{1}{2}\hat{V}\hat{I}\cos\phi \quad \text{and} \quad P_{dc} = \hat{V}\hat{I} \tag{1.6}$$

where

$\hat{V}$, $\hat{I}$ – peak transmitted voltage and current respectively,

$\cos\phi$ – power factor of the Grid network

($\simeq 0.8$ lagging for the United Kingdom)

It is seen that a wind farm with a dc cable link carries twice the power for the same installation cost.[14]

[12] Apart from its scheduled annual overall and an insignificant number of short duration trips, its actual capacity factor was 60%.

[13] By using forced draught cooling towers, for example.

[14] With ac transmission, the above P is the average power per phase per cycle (i.e., in one conductor). However, three-phase cables with a constant instantaneous power compare even less favorably as the peak voltage between phases is $\sqrt{3}\hat{V}$.

Table 1.4
Annual Wind-Power Data 2005–07

Country	Spain	Denmark	Germany	United Kingdom	United States
Capacity factor (%)	24.6	24	16	28	16–20
MCR wind power (GW)	41	4	81	5[a]	105

[a]After commissioning Isle of Thanet, 2010.

The stochastic meanderings of high and low barometric pressure zones across the planet are patently beyond human control. Also a high-pressure region, in which there is little or no wind, can blanket a sizeable portion of a European country for as long as a week thereby disrupting electric power generation. Correspondingly reduced annual capacity factors and installed wind powers [36,37] are illustrated in Table 1.4 for 2005–07.
Though large countries like the United States and Russia can "hedge" by spatially distributing their wind farms, smaller Northern European nations can face major disruptions which are especially critical during their peak winter demand periods. Accordingly, if wind power is to enable a significant reduction in European carbon emissions, a solution to its intrinsic intermittency must be found in order to preserve the security of national power supplies. Due to the withdrawal of government support from the Severn Barrage scheme, the remaining option for a materially sized renewable UK energy resource is wind power, which is potentially required to be the largest in Europe [42].

Though electrochemical batteries or diesel generators are reliable backups for isolated low-power applications, they are totally nonviable for sizeable offshore wind farms in a national Grid. For instance wind power developments in the United Kingdom have installed, either in construction or in planning a rated 18 GW for operation by 2020. Because this power equates to the output of about 18 large fossil or nuclear stations, radical measures are necessary to secure the national power supply. Toward this end a memorandum of understanding [38,39] was signed in 2009 for the creation of a European Supergrid network at an estimated cost of €30,000M, but construction plans remained under discussion in 2012. In particular, the French–UK ac connector of 2 GW is to be supplemented by dc links from both Norway and the Netherlands.[15] However, if a shortfall in UK power is drawn from a continental connector, it would be at a premium price: especially in mid-winter. Furthermore, though Norway has an installed hydro-capacity [43] of about 30 GW, its peak demand in winter is around 22 GW, so this country could not offset [44] a major meteorologically induced disruption to the projected UK's wind power generation. On the other hand, because reliable predictions of high barometric pressure zones can now be made several days in advance, UK-sited fossil and nuclear stations would be able to increase their outputs at demanded rates well within operational constraint limits.[16] Because nuclear power plants are capital-cost dominant and fossil stations are fuel-cost dominant, it is cost-effective for nuclear stations to supply the largely predictable daily base load, and for fossil stations to supply the more rapidly varying load excursions. For this latter purpose a number of fossil stations operate at around 80% of nameplate ratings (MCR) to provide a so-called "spinning reserve." Sudden very rapid demands such as the unexpected disconnection of a large generating unit or a pause in a very popular TV program are also buffered by the pumped storage schemes at Dinorwic [40] (1.8 GW), Ffestiniog (0.36 GW) and Ben Cruachan (0.44 GW) but due to the very special topography required it is unlikely that other such suitable UK sites can be found. It is to be concluded that a "mix" of wind, fossil and nuclear stations has become necessary for a flexible, secure and economical UK power supply.

Total capitalization of the Isle of Thanet wind farm [31] is $1353 M or $4.5 M per installed MW. However, in addition to the loss of revenue from an inevitable shortfall of delivered power, there are the presently uncertain capital and operational costs of the necessary backup systems [52]. These depend on the future chosen "mix" of fossil and nuclear plants together with charges levied for power dispatched over the as yet unbuilt European Supergrid. Currently quoted costs for wind generation are therefore subject to considerable uncertainty, which perhaps led the Royal Dutch Shell Company to withdraw its support from renewable energy schemes [45]. On the other hand the UK Sustainable Development Commission [46] reports that "the economics of nuclear new-build are uncertain," but this statement is contradicted by decades of

[15] To access existing grid connections between Germany, France, Belgium, etc.

[16] See Chapter 3.

worldwide practical experience—especially in the Far East. It is possibly of note that the UK's coalition government disbanded [47] this Commission in July 2010.

Finally, the capacity factor of UK wind turbines in Table 1.4 appears heavily biased: possibly toward offshore systems. According to the UK regulator OFGEM that for the nominal 2 MW Reading city installation was just 15.4% during 2010. Even more significantly the market value of its electricity production was £0.1M, but thanks to a government subsidy, its owner Ecoelectricity received £0.13M. Under these conditions, large-scale wind power is an excellent investment for UK utility companies! Clearly, the true cost per actual MW generated, as well as environmental impacts [324] and Grid compatibility should be properly considered when deciding the future "mix" of UK generating plant.

1.7 FOSSIL-FIRED POWER GENERATION

The global industrial revolution in the late nineteenth century originated in the United Kingdom with coal powering steam engines and iron smelting. Now in the twenty-first century electricity generation by coal is constrained by economics, carbon emission penalties and the availability of cleaner natural gas [49]. Though coal remains the planet's largest fossil-fuel resource [55] its large-scale utilization is presently incompatible with the pursuit of low carbon emissions. As well as carbon dioxide, other environmentally damaging combustion products [55] include sulfurous oxides and fly ash which contains mercury, arsenic and radioactive uranium and thorium. In fact without fly ash capture equipment, coal-fired stations would contribute significantly to background radiation. Table 1.5 shows the estimated coal

Table 1.5
Coal Resources in 2006–07

Country	Australia	China	Germany	India	United Kingdom	United States
Total coal resource (G tonne)	600	1438	246	81	190	2570

reserves [53] of all types[17] for a number of industrialized countries in 2006–07. It suggests the strong motivation [50] to develop so-called Clean Coal technology for reducing pollutants and achieving fuller combustion by pulverization. Because nitrogenous oxides are produced at combustion temperatures above 1370 °C, temperature control between 760 to 927 °C eliminates these without the need for flue-gas scrubbers [50].

When finely divided limestone is intermixed with pulverized coal, 95% of the sulfurous precursors of acid rain are absorbed: but at the expense of larger carbon dioxide emissions. Present research searches for more suitable sorbants [50] and Carbon Capture processes [54]. By measuring the ratios of stable isotopes of carbon dioxide and noble gases, recent studies of nine gas fields in North America, China and Europe have established that underground water is the principal sink and has been so for millennia [54]. These experiments could provide a basis for validating mathematical models of future storage locations and for tracing captured carbon dioxide. However, on-going capture tests require 25% of the Longannet 2.4 GW station's output [56] so that an economically viable process has yet to be developed. A sum of £1 billion was allocated for this purpose in the October 2010-UK Spending Review, but was declined by a consortium of Scottish Power, Shell, and National Grid.

Commercial quantities of natural gas were discovered in the North Sea during 1965, and since then in many other countries. Combined cycle gas turbine plants (CCGT) [51] have subsequently had a material impact on new-build generating capacity as illustrated [57] for the United Kingdom in Figure 1.2. In these, gas first powers a gas turbine whose exhaust via a heat exchanger provides steam for a conventional steam turbine with feed water heating and reheat to enhance thermal efficiency. With the lower cost CCGT configuration, an alternator is driven by gas and steam turbines sharing a common shaft, while with the more flexible but more expensive multishaft arrangement, each has its own alternator. Typical burnt-gas and steam-inlet temperatures for CCGT and coal fired plants are

$$T_{CCGT} \simeq 1000\,^{\circ}\mathrm{C} \quad \text{and} \quad T_{\mathrm{coal}} \simeq 570\,^{\circ}\mathrm{C} \tag{1.7}$$

[17] Anthracite, coking, lignite, and steam.

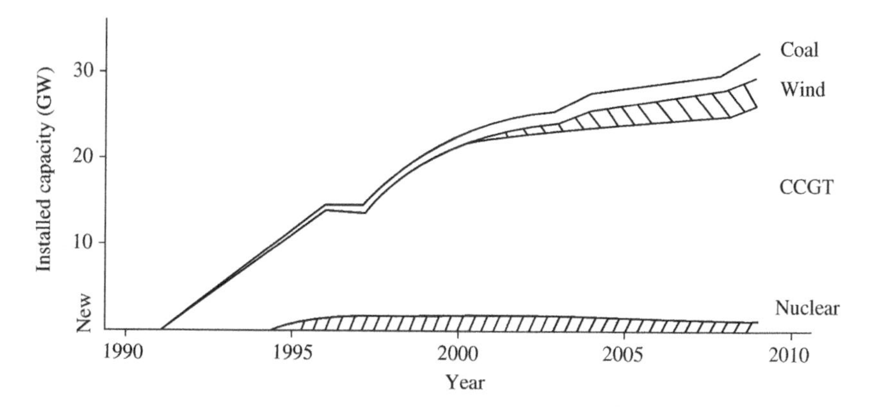

Figure 1.2 Illustrating the Cumulative Investment in UK Generating Plant [57]

For an ambient condenser temperature of 30 °C, the corresponding Carnot Efficiencies are derived from equation (1.2) as

$$\eta^*_{\mathrm{CCGT}} \simeq 76\% \quad \text{and} \quad \eta^*_{\mathrm{coal}} \simeq 64\% \tag{1.8}$$

but due to thermodynamic irreversibilities, the practical efficiencies achieved are

$$\eta_{\mathrm{CCGT}} \simeq 50\% \quad \text{and} \quad \eta_{\mathrm{coal}} \lesssim 40\% \tag{1.9}$$

The preferential installation of CCGT units shown in Figure 1.2 is now clear. Because investment in a privatized market is governed by a commensurate return on capital expenditure and associated risks, and because UK electricity prices are set by those for CCGT generation [52] and infrastructure provision, utility companies are assured a fair and timely low-risk return. During 2011 CCGT stations delivered around 44% of the UK electricity: but what of the future?

With the reduction in North Sea gas production, the United Kingdom has now become a net importer and is therefore potentially beholden to the vagaries of international markets or the political whims of some exporters. Accordingly coal-bed methane, shale and conventionally drilled gas production are being actively investigated to regain a self-sufficient supply. Hydraulic fracturing [323] or "fracking" is particularly successful in speeding up gas flow rates from shale or other "tight" reservoirs to render them economical. This technology involves unconventional horizontal drilling along a promising shale strata followed by the injection of high-pressure water and chemicals. The process has transformed US gas production from next to zero in 2000 to an almost self-sufficient 13.4 billion cubic feet per day. Cuadrilla Resources plc claims to have discovered a potential 200 trillion cubic-feet shale gas reservoir in the northwest of England, and the British Geological Survey suggests a total onshore resource of some 1000 trillion cubic-feet. However, test drillings have elicited small earth tremors[18] at Blackpool and there are further concerns regarding the contamination of drinking water supplies. Consequently commercial development has been halted until the Department of Energy and Climate Change has completed a review. Even if an abundance of onshore gas becomes available, a detailed study [57] reveals that CCGT generation alone could not fill the UK energy gap within the ratified carbon emission targets [1,52], so that a nuclear component appears as the necessary reliable complement in the eventual "mix" of generating stations. To achieve an economical fuel cycle (burn-up) and the intervention of safety circuits nuclear stations must supply the more slowly varying and largely predictable national base load.[19] Accordingly CCGT plants with preferably Lamont boilers [117] are better able to provide the more flexible and faster responses to rapid unscheduled changes in Grid power demand.

Sizeable UK oil-fired power stations like Poole and Marchwood were decommissioned over 10 years ago, but a number of small ($\lesssim$10 MW) units still exist to buffer unexpected peaks in national demand. These relatively low thermal efficiency, but highly responsive "peak lopping" units presently contribute around 1% of national energy consumption [49].

1.8 NUCLEAR GENERATION AND REACTOR CHOICE

Natural uranium consists for the most part of very weakly fissile U-238 and 0.7% of highly fissile U-235. In broad statistical terms, neutrons in the natural substance very largely encounter U-238 atoms which

[18] Refer to the Basel CH experience in Section 1.2.

[19] See Figure 1.1.

usually slow them down. When neutronic energies are reduced to between 0.1 and 1.0 eV they are captured by the pronounced resonance absorption bands [58] of U-238 nuclei, which together with surface leakage obstructs a self-sustaining chain reaction in the natural material.

Thermal[20] nuclear reactors achieve a self-sustaining reaction by embedding uranium oxide[21] fuel rods in a carefully contrived matrix of moderating material. Collisions with moderator nuclei reduce neutronic energies to below the U-238 absorption bands in a short distance after fission, so thereafter fission with very largely U-235 atoms takes place to create a self-sustaining reaction with 32 pJ of heat released per fission. Moderators are usually constructed from graphite (AGR), heavy water (CANDU), light water (PWR, BWR) or a composite of any two of these (RMBK, SGHWR). The fissile concentration (enrichment) is often increased radially to an average of around 3% in order to achieve a more uniform and therefore more cost-effective power production. Unlike heavy water, light water is both an effective moderator and absorber [58], so its conversion to less dense steam reduces both neutron moderation and absorption. By astute design of core-lattice geometry and fuel enrichment, light water reactors outside Russia have always been designed to become under-moderated with increasing steam production [61], and this policy is vindicated by the Chernobyl disaster [12]. Indeed a negative power reactivity coefficient is now a necessary prerequisite for licensing by European Regulatory Authorities.

A self-sustaining chain reaction is achieved in a fast reactor by ensuring that the average neutronic energy remains well above the absorption resonances of U-238. Inelastic scattering and parasitic absorption of fast neutrons are overcome by a typically 20% enrichment with a mix of U-235 and Pu-239 oxides. The power density (W/m^3) is consequently so large that individual fuel pins can have only small diameters ($\simeq$ 5 mm). To avoid a significant moderation of neutronic energies these pins are closely spaced in hexagonal subassemblies and cooled by liquid sodium. Though parasitic absorption in structures and fission products (e.g., Xe-135) is relatively less in fast reactors, damage mechanisms are clearly aggravated and careful choices of

[20] So-called because fissions largely occur at neutron energies around the thermal vibrational energies of the fuel molecules.

[21] Allows a higher fuel temperature than the metal and a longer life.

materials are necessary. For example, liquid sodium leaches out carbon from stainless steel, so this fuel cladding must be niobium stabilized. Despite higher fuel fabrication costs than a thermal reactor, research has doubled its in-service life and capital installation costs are somewhat offset by a smaller reactor core. Uranium for the Russian nuclear program and Skandia for rocket motor exhausts were found in a very remote region around Aktau. Power, desalinated water and fish (farmed sturgeon[22]) for the mining complex were provided during 1973–94 by the BN350 fast reactor: chosen perhaps for the easier transport of its relatively smaller core. The unique advantage of fast reactors resides in their better "neutron economy" which enables the production of more fissionable material than that consumed. For this purpose the core of highly enriched subassemblies is surrounded by breeding blankets of natural uranium to give

$$\text{U-238} + 1\,\text{neutron} \rightarrow \text{Pu-239}\,\text{fissile} \tag{1.10}$$

If the existing stock of UK nuclear materials were to be used in this way, the estimated energy would be comparable with recoverable coal reserves [60]. However, with uranium now so plentiful, economics and strategy favor the construction of thermal reactors, whose choice is now considered.

The engineered slowing down of neutrons from around 2 MeV at fission to thermalization at 0.025 eV is almost entirely done by elastic collisions with moderator nuclei. Applying the law of conservation of momentum and assuming spherically symmetric scattering in a center of mass coordinates the mean logarithmic decrement ξ of neutronic energy per collision is given by [58,61]

$$\xi = 1 + \left(\frac{\alpha}{1-\alpha}\right)\log\alpha \quad \text{with} \quad \alpha \triangleq (A - 1/A + 1)^2 \tag{1.11}$$

where A is the mass number of moderator nuclei. A logarithmic compression is appropriate by virtue of the wide energy range involved. Though an effective moderator has a large ξ, the probability of colliding with a nucleus must also be large. Hence, the slowing down power of moderator is defined as $\xi\sum_s$ where $\sum_s$ is its macroscopic scattering

[22] Using the warm condenser outflow into the Caspian Sea.

Table 1.6
Properties of Common Moderators [58,61]

Moderator	$\sum_a(\text{cm}^{-1})$	$\sum_s(\text{cm}^{-1})$	ξ	$\xi\sum_s$	$\mathcal{M}_R$
Light water	0.022	3.45	1.0	3.45	157
Heavy water	0.00003	0.45	0.57	0.26	8667
Graphite	0.00026	0.39	0.16	0.06	240

cross section [58]. In physical terms $\xi\sum_s$ represents the statistically expected energy loss in a 1 cm cube of a moderator. Moderators also absorb neutrons which incurs an increased fuel enrichment, so their efficacy is characterized by [58]

$$\text{Moderating Ratio}\,(\mathcal{M}_R) \triangleq \xi\sum\nolimits_s \Big/ \sum\nolimits_a \tag{1.12}$$

where $\sum_a$ is the moderator's macroscopic absorption cross section. Values of the above parameters for commonly employed moderators are shown in Table 1.6, which reveals in terms of $\xi\sum_s$ the relative compactness of light water cooled and moderated BWRs and PWRs. Because civil engineering costs dominate nuclear power plant construction,[23] and because a slightly increased fuel enrichment to offset additional neutron absorptions is cost-effective, private utilities worldwide favor these designs. Though both types have capacity factors of around 90% [25], other features determine their relative merits.

A schematic diagram [62] of a PWR plant is shown in Figure 1.3 with details of an actual reactor [205] in Figure 1.4. Electric heaters, cooling coils and sprays within a pressurizer produce and control the size of a steam bubble to set the primary circuit pressure at 15.5 MPa. Because the primary coolant remains single-phase, the "elasticity" of this bubble accommodates its thermal expansion or contraction. Under fault conditions an electromagnetically operated valve with manual backup enables steam venting into the steel-lined reinforced concrete containment, where cold sprays would dissolve fission products (e.g., iodides) and prevent its over-pressurization. In addition there are recombiners to mitigate potential detonations of hydrogen caused by oxidation of the fuel elements. The primary circuit of a BWR directly

[23] See Table 1.7 below.

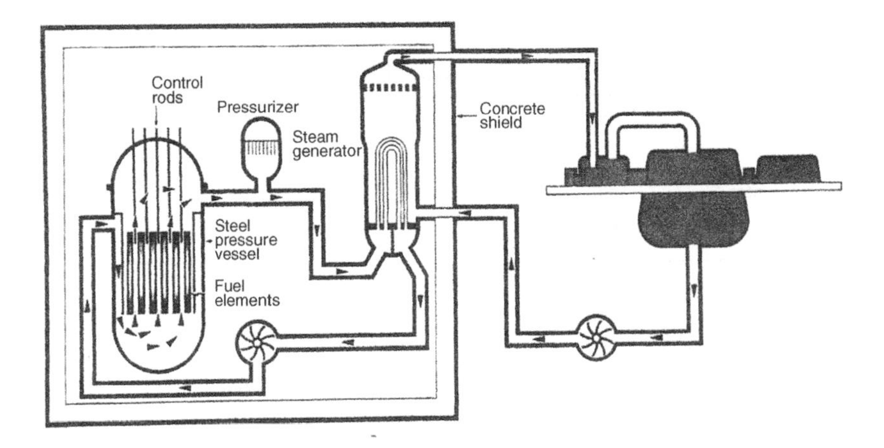

Figure 1.3 A Schematic Diagram of a PWR Plant [66]

supplies the saturated steam component of its 2–phase output [62] to steam turbines. By allowing water to boil at the lower pressure of 7–2 MPa, a BWR needs neither steam generators nor a pressurizer. Though the lower operating pressure appears to further reduce initial capital cost by way of thinner pressure vessels, core dimensions must be increased for the lower linear fuel rating (W/m) which is necessary to prevent a damaging dryout transition from nucleate boiling [63,64]. Also as described in Chapter 3, steam in boiling subchannels can give rise to flow instability which is another potential source of fuel damage. Furthermore, power control in a BWR is patently complicated by interactions between internal steam volumes, coolant flow rate and the insertions of its cruciform control rods. Despite the additional costs of components and a thicker pressure vessel to inhibit nucleate boiling at 15.5 MPa, PWRs are more generally favored as they are cost-effective in avoiding the above design and other operational or safety issues. Specifically, in the rare event of fuel melting, the separate steam-generator units of a PWR provide an excellent heat sink[24] and additional isolation between the reactor and its environment. Under these circumstances, secondary-side steam vented to atmosphere in an accident

[24] A residual heat removal heat exchanger is also provided.

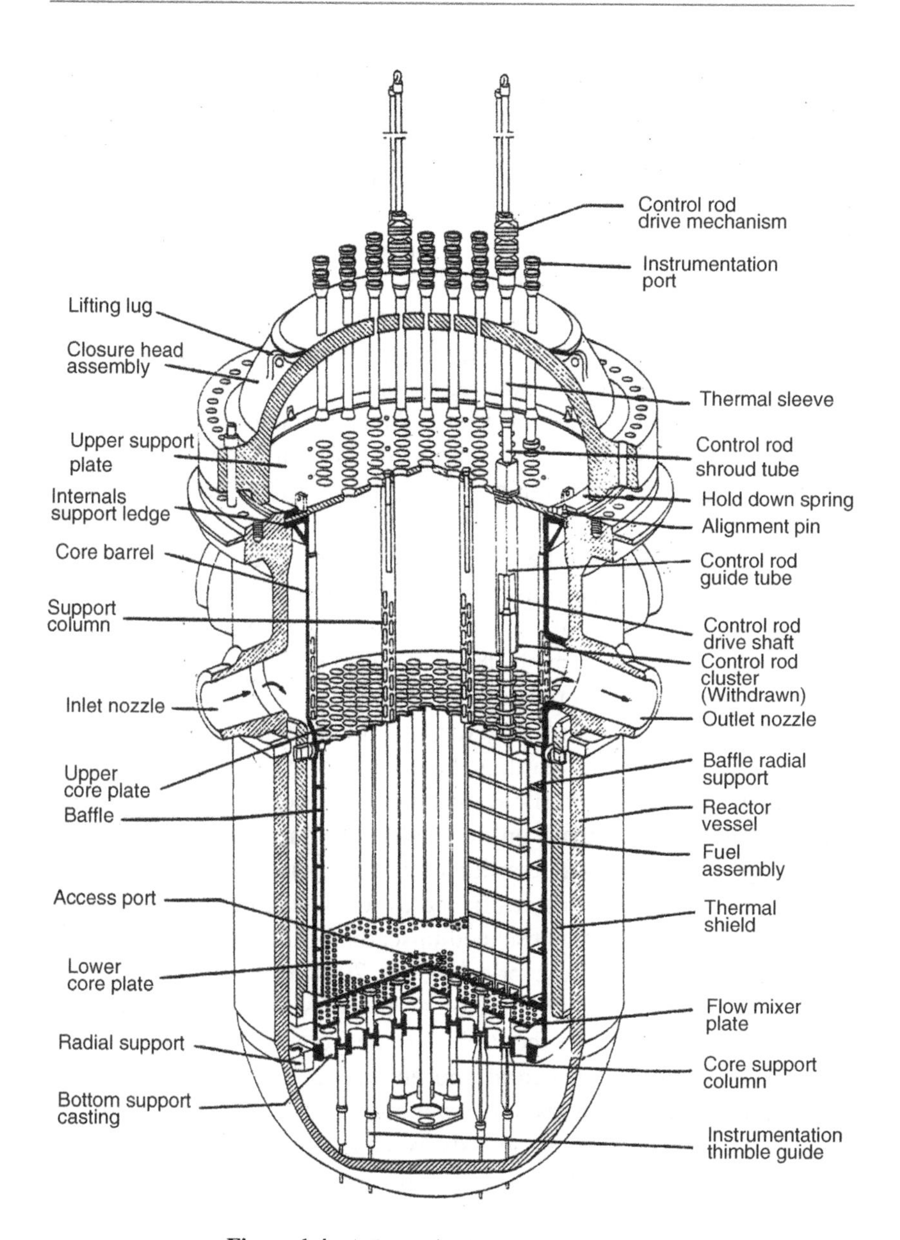

Figure 1.4 A Pressurized Water Reactor [178]

"bleed and feed strategy" would be far less radioactive than that released from a BWR.

Public confidence in nuclear power was shattered by the reactor incidents at Three Mile Island [66] (1979) and Chernobyl [12] (1986). However, the former galvanized the start of globally intensive safety research,[25] as well as new stringent operating legislation and decommissioning techniques [69]. In this respect the author's own research at AEEW was abruptly shifted from intact plant control studies to investigations of explosive boiling and its potential damage to internal reactor structures. Others conducted theoretical and experimental studies [67,68] into the impact of a jet fighter (Tornado) on a reinforced concrete reactor building, or a dropped flask of highly radioactive reactor fuel, or the detection before their critical length [96] of embrittlement cracks in PWR pressure vessels. By 1992 international research had confirmed the effectiveness of active accident control measures to mitigate the consequences of fuel melting in both fast and thermal reactors. This research still continues into the design of passive emergency cooling systems [108–110] based on natural circulation to avoid auxiliary power supplies.

While this book was being written, the northeast coast of Japan experienced horrendous earthquakes, tsunami and a major nuclear incident at the Fukushima BWR plants. Some brief personal comments here concerning its impact on nuclear safety and future construction appear apposite. National licensing of nuclear plant operations requires a demonstrable engineered resilience to local seismic activity. In this respect all Japanese plants escaped unscathed from the earthquakes in 2009 and 2010. Also just prior to the 2011 tsunami, an effective neutronic shutdown was effected on all the West Coast and East Coast Fukushima plants: despite the Richter-scale 9 quake. Media reports and pictures indicate that structural and emergency core-cooling systems failed at Fukushima as a result of swamping by the subsequent tsunami whose estimated 14 m height grossly exceeded the design limit of 5.3 m. Historic data on tsunamis is therefore less complete than for their earthquake precursors, and flood defenses on the East Coast clearly proved inadequate. With hindsight, Japanese nuclear stations should have been built on the West Coast which is sheltered by mainland China. It is considered here that the resulting opposition to nuclear power

[25] Coordinated by the IAEA of the United Nations Organization.

station construction in locations not threatened by tsunami or serious flooding is an over-reaction. Likewise, whilst inadequate UK planning has allowed homes to be built on flood plains, this is no reason to refuse their construction elswhere on suitable sites. Also it is widely believed that diagnostic X-rays are our only exposure to radiation and that any exposure materially damages our health. In fact cosmic rays and natural radioactive decay in the earth's crust cause a continuous pandemic exposure[26] and though citizens in the granite city of Aberdeen receive about three times the background exposure of Londoners, no statistically significant increase in pertinent cancer cases occurs. Though I–131[27] ingressed into Tokyo's drinking water from Fukushima it was around only 1/5th the UK's safe limit for all ages, and the Japanese advocated consumption by adults only. Moreover, very conservative exclusion zones and contamination limits on farm produce were imposed. However, it was the admission of falsified plant safety reports [170] in February 2010 that created the material loss of public confidence.

During 1993 numerous early retirements of UKAEA professional staff left a minority to develop successful decommissioning and waste glassification technologies that are now deployed worldwide by AMEC plc [70]. At Winfrith the Zero Energy Facilities and the High Temperature Dragon Reactor have now been properly decommissioned, and the buildings of the SGHWR demolished. However, water reactors themselves take relatively longer due to their thicker corrosion deposits which hold up larger quantities of radionuclides. Nevertheless progress to date suggests that the entire site will be outside nuclear regulations during 2039–48. While the storage of glassified radioactive waste raises public concern in some quarters, the Swedish towns of Forsmark and Oskarshamn actually competed [320] for a high-level waste facility to be built in their respective neighborhoods [72]. Construction at Oskarshamn was approved in 2009 with a start date in 2013, and on completion concreted waste in 25 tonne copper-sheathed stainless steel drums embedded in impervious Bentonite clay cushions will be buried in stable igneous rock tunnels. A further safeguard is that these containers are to be retrievable for inspection because the waste itself has potential industrial or medical applications. Unlike the environmental release at Bhopal of dioxin with an indeterminate active life, the

[26] About 100 mrem = 1 mSV.

[27] Half-life 8 days.

radioactivity in nuclear waste reduces to that of mined ores in about 7000 years [321]. The natural fission reaction at Oklo some 1800 million years ago provides evidence that igneous rock formations alone can contain fission products for well over this period.

Since the Three Mile Island incident in 1979 the worldwide deployment [73] of 269 PWRs has operated at high-capacity factors [25] and without a major failure of the nuclear technology itself. A contributing factor is the shared experience within the PWR Operators Club that has led to safety-enhancing retrofits and procedures. It is therefore contended that PWRs offer a technically sound and safe solution to an impending electric power deficit. Vindication of this strategy is further offered by the willingness of populations around existing AGR sites to accept replacement PWRs: particularly when endorsed by the families of local plant staff. However, technical and safety issues alone are insufficient for renewing the UK nuclear power program in the now privatized electricity industry. Specifically, the huge capital investment must be largely met by private equity rather than as previously from public funds. In this context a commensurate return on shareholder funds must be incipiently visible, and towards this end an appropriate financial framework must be pre-established by the government and its regulator OFGEM. Some pertinent factors for consideration are now described.

The lifetime cost breakdown [74] for new PWR plant is shown in Table 1.7 which reveals the largely dominant construction cost. Build-times, and therefore costs, vary between 4 to 7 years depending on national working practices and the number of repeat orders. No revenue is evidently forthcoming during construction, but capacity factors once operational are as high [25] as 90% over a design life [74] of 60 years with insignificant carbon emissions and highly efficient land use (see Section 1.6). However, due to the initially high capital expenditure and delayed revenue, an estimated payback period of 30 years is required [52]. On the other hand as explained in Section 1.7, CCGT generation

Table 1.7
Lifetime Percentage Costs of New-Build Nuclear [74]

Cost	Construction	Fuel	Decommissioning	Waste Management
%	60–70	17	5	10

is radically different and more favorable to private equity investment by virtue of its timely income and assured profit. Though nuclear appears to be the lowest cost source of low-carbon electricity generation [52], it has to compete under the present UK regulatory framework with the base-load costs of CCGT units, even though these are not compliant with the mandatory 2020 UK emission targets. Indeed the Chief Executive Officer of RWE Nuclear argues [75] that the government's renewable obligations tariff should be changed to a low carbon obligations tariff in order to fairly characterize the role of nuclear power. The financial similarity between wind and nuclear power investment clearly supports his argument. However, levelized costs for new-build nuclear are estimated [52] as $92 to $123 per MWh, which are well below[28] $246 to $308 per MWh for offshore wind turbines: even before the cost of the necessary backup systems is included.

The first thermal reactor for commercial electricity was completed in 1956 at Calder Hall. Despite its description as "The Peaceful Use of Atomic Energy," there remains public apprehension that commercial nuclear power stations are also sources of weapons material. In essence, nuclear weapons create an explosive growth of the neutron population in a mass of largely fissile material as an end in itself, or as the initiator for a fusion device. For the Hiroshima A-bomb an appropriate mass of uranium was highly enriched with U-235 to restrict parasitic absorptions by U-238. The Nagasaki weapon was designed around plutonium recovered from a specially contrived fuel cycle which ensured very high concentrations of fissile Pu-239 relative to that of the Pu-240 created by another neutron absorption. Because the higher mass isotope is an unstable α-emitter [76], a sufficiently high concentration would induce the partial triggering of a plutonium-based weapon. Accordingly weapons-grade plutonium has a specified Pu-240 concentration of less than 7%. During the in-service life of thermal reactor fuel, fission of created Pu-239 forms a partial and immediate replacement for "consumed" U-235 atoms, but burn-up of Pu-240 proceeds at a slower rate. Consequently the relative concentration of Pu-240 increases with increasing fuel burnup. Fuel pins for power reactors are precisely engineered fabrications that embody years of research to enhance safety and the economy of electricity generation. Burn-up targets[29]

[28] At $1 = £0.65.

[29] Presently over 60 MWd per kg of U in pristine PWR fuel – Ref. [77].

for commercial power reactors have always been determined by these considerations so that Pu-240 concentrations in recovered plutonium are too high for weapons purposes. Nuclear power generation has been and is therefore still divorced from nuclear weapons.

1.9 A PROLOGUE

Though esoteric control engineering theory might be outside some reader's interest, a brief overview of Nyquist's [131] and Rosenbrock's [123,134] theorems is necessary in order to appreciate the practical problems described in Chapters 2 and 3. Here the stabilities of neutron reactor kinetics, flow in boiling channels and a national electricity Grid are analyzed. At first sight it appears surprising that a national Grid stability criterion [80,84] can be formulated as a single input–single output problem when many plant controls and a myriad of ac generators and motors are intimately involved. This example particularly illustrates the important engineering skill of identifying the reduced set of variables that dominate a complex physical system in order to effect a successful solution.

Subsequent chapters concern some research activities and operational legislation that aim to underwrite the safety of nuclear power plant following the Three Mile Island incident [66] in 1979. In this respect international collaboration has been sponsored by individual governments, the OECD, the European Union[30] and the PWR Operators Club. The success of these initiatives is confirmed by the absence of any later major technological or operational faults[31] with BWRs and PWRs.

Chapter 4 cites some European statutory probabilities [59,65] for the occurrence of Design Base and Severe Accidents (fuel melting) along with operational requirements [59,108] relating to the progression of plant damage and the need for an operator command structure based on professional skills and training. Hazards and Risks from some pertinent fission products after an unlikely environmental release are described, together with the legislated exposure limits (Sv) for on-site operators and the neighboring public. Though Event Trees and

[30] See Ref. [108] for details of European Utility requirements.

[31] Section 1.8 contends that Fukushima is a site-planning error.

Risk analyses are increasingly deployed for safety assessments of many industrial processes, weaknesses in their application are identified as failures to address benefits and humanity's greater acceptance of one manner of death from another. Nevertheless with no real alternative they are still used for assessing nuclear power plant, whose principal Risk to people is the development of thyroid cancers due to the natural concentration of absorbed radioiodides [163] after an unlikely environmental release of fission products. In this context, it should be noted that 80 to 90% of naturally presenting cases are successfully treated by surgery [164]. The Farmer–Beattie analysis, which is repeated here with the more appropriate Poissonian rather than a Normal distribution, demonstrates that the expected annual incidence of thyroid cancers from the spectrum of Severe Accidents in an AGR station is some two orders of magnitude less than the number of natural presentations. Granted rationality, nuclear power with its attendant benefits should therefore have gained full public approval! Accident precursors and safety systems for fast reactors and PWRs are briefly outlined. However, because the latter appears to be the most widely adopted type of future civilian plant, the robustness and diversity of their safety systems is illustrated for a large loss of coolant accident (LLOCA).

There are rare circumstances when a coherent explosive rate of heat transfer occurs as a result of a high-temperature liquid mixing with a readily vaporized one. Though unimaginably powerful natural explosions took place at Krakatau and Santorini, the situation of Severe Accidents in PWRs with molten core debris (corium) is radically different. Specifically, the natural events involved a rapid violent mixing of Gigatonne quantities, whereas the spatial neutronics and hydraulics of water reactors allow only the progressive melting of their 100 tonne cores over hours [59,65]: and then only if all the diverse safety systems were to fail concomitantly. In nuclear safety assessments such an explosive heat transfer is called a molten fuel–coolant interaction (MFCI). The Three Mile Island incident in 1979 invigorated global research into the progression of core damage [59,65,213,269,270] and the physical phenomena in Severe Accidents. In this respect, the bounding Hicks–Menzies analysis of 1965 had raised concern that not unrealistic masses of corium in an MFCI could challenge the integrity of a fast or water reactor's containing vessel. As well as releasing radioactive fission products, its rupture could create rapidly accelerating metal fragments (missiles) which could potentially breach the surrounding reinforced concrete containment building, and thereby allow an environmental release of radioactivity. However, a simulation developed by the author during 1989–92 confirmed the 4 to 5% conversion efficiency of heat into mechanical work that had been observed in many independent kilogram-sized MFCI experiments. This efficiency is some six times smaller than an isentropic Hicks–Menzies value which disregards the highly effective anisentropic heat transfer from an MFCI bubble into the bulk coolant by interfacial condensation. Identification of this physical process allows a valid extrapolation of the 4 to 5% experimental-scale value to tonne-sized reactor quantities thereby materially benefiting reactor safety assessments. Novel finite element models for the impact of plant missiles or aircraft on reinforced concrete structures or major pipe work were also validated during this same period. Following the foreclosure of the EFR project, European research on MFCI and impacts was discontinued. The pertinent UK reports were then archived leaving a much smaller staff complement to pursue successful reactor decommissioning, waste glassification and passive safety systems. Chapters 5 and 6 respectively outline these MFCI and impact research archives to assist engineers and scientists newly entering the resurgent nuclear industries.

After the Three Mile Island incident the nuclear power industry sought unremittingly to improve in-depth plant safety [298,302]. For example, more robust fuel cladding [300] and steel [276,277] or concrete containments [286] have been developed. In addition cooling circuits deploying natural circulation have been proposed [108] as potentially more reliable and cost-effective by eliminating the need for active power supplies and pumps. However, due to intrinsically smaller heat transfer rates than with forced convection, natural circulation cooling systems are relatively much larger. As about 60 to 70% of capital costs reside in civil engineering works, they do not appear economically viable for the main on-load cooling of water reactors of order 1 GWe. Nevertheless, because the radioactive decay of fission products peaks at around 10% of pre-trip (scram) power, passive safety circuits exploiting heat removal by natural circulation are cost-effective [108] and their development has become an ongoing activity [109] associated with proposed Generation IV plants [109,298,302]. Chapter 7 concludes the book with a discussion of further advantages and disadvantages of heat removal by natural convection and a passivity classification for reactor safety systems. All proposed passive safety systems address the problems of decay heat removal to ensure core debris-bed cooling [65], pressure relief inside a concrete containment [101] and the blocking of emergency core cooling systems (ECCS) by debris [100]. Because the driving forces from differential densities and gravity are relatively much weaker than with forced convection, careful design and validated analyses are necessary to be sure that these passive systems function as intended. Indeed the IAEA falls short of recommending them as direct replacements for the active safety systems in presently operational plants [10].

A final thought to ponder is that the human and environmental consequences of the Three Mile Island, Chernobyl and Fukushima nuclear incidents are all together dwarfed by the 171,000 deaths caused by the failure of the Banqiao hydro-dam [11]. It is hoped that Chapter 1 will contribute to an objective quantified debate on future electrical energy generation which encompasses national and global issues of Grid demand patterns and land resources [324].

CHAPTER 2

Adequacy of Linear Models and Nuclear Reactor Dynamics

2.1 LINEAR MODELS, STABILITY, AND NYQUIST THEOREMS

All forms of commercial power generation involve the controlled manipulation of plant variables to achieve a prescribed contribution to national demand. In fossil and nuclear plants the heat source, boiler feed pumps and turbine control valves are the pertinent items. For wind turbines the rotor-blade angle and generator excitation are the relevant quantities. When several plant variables require control, engineers describe the situation as a multi-input multi-output (MIMO) problem. On the other hand car speed control via fuel-injection rate exemplifies a single-input single-output (SISO) problem. Even over their intact operating regimes fossil and nuclear power plants are materially non-linear[1] and distributed.[2] In this framework control design cannot be implemented analytically by existing mathematics. However, for sufficiently small perturbations about a given operating point, plant

[1] Consider typical heat transfer correlations; see Refs [63,64,143,219].

[2] Necessarily described by partial differential equations.

Nuclear Electric Power: Safety, Operation, and Control Aspects, First Edition.
J. Brian Knowles.
© 2014 John Wiley & Sons, Inc. Published 2014 by John Wiley & Sons, Inc.

dynamics can be approximated by a finite set of ordinary linear differential equations that enable methodical insight into the effects of negative feedback (control), variable interaction, and stabilization by means of linear compensating algorithms [79,124]. Aizerman [77,78] conjectures that an ordinary non-linear differential equation that has small signal (linearized) stability about every steady state also has global stability [77,126]. Though no general analytical proof of this exists, it forms a basis for the successful control of fossil and nuclear power plants.

First of all a number of linear models are derived that reasonably characterize plant parameter changes over the normal operating regime. Experience indicates that steps of about 10% of the maximum continuous rating (MCR) are usually sufficient for the purpose. After engineering linear stabilizing algorithms for each power level, a compromise is generally sought that ensures adequate stability (transient damping) for all. This problem is eased for power plants because unlike defense equipment speed of response is not the priority. With no certainty that normal maneuvres can be accomplished, confirmation is imperative using a detailed non-linear plant simulation [117,141] whose individual models have been validated as far as possible against existing plant items. This same non-linear simulation can also provide the required linear models as illustrated by examples in Chapter 3.

Linear control system theory is often couched in the abstract algebras of finite dimensional Banach or Hilbert Spaces [110–112]. However, for engineering design purposes linearized plant dynamics are specified in the state equation format [122,123,134]

$$\dot{x} = Ax + Bu \quad \text{and} \quad y = Cx + Du \tag{2.1}$$

where

$x(t)$—a state vector containing a finite number (n) of Laplace transformable functions

$\dot{x}$—temporal derivative of x

A, B, C, D—real matrices with respect to a convenient Cartesian coordinate system

$u(t)$—an input vector

$y(t)$—the output response vector

Defining the Laplace transform [119,120] of a vector of time function as that of each of its components, then from equation (2.1)

$$\bar{x} = (xI - A)^{-1}B\bar{u} + (sI - A)^{-1}x_o \quad \text{and} \quad \bar{y} = C\bar{x} + D\bar{u} \qquad (2.2)$$

where $\bar{x}$ denotes the Laplace transform of $x(t)$ etc. and x_o its initial value. In the present finite dimensional context the spectrum [110,111,121] of an arbitrary matrix L corresponds to values of s for which

$$\det(sI - L) = 0 \qquad (2.3)$$

and these roots are termed the eigenvalues of L. A determinant and eigenvalue spectrum are properties of the underlying linear mapping and are independent of the chosen Cartesian coordinate system. Equation (2.3) defines the characteristic polynomial [110,111,121], and for a real transition matrix A it has real coefficients, so the eigenvalues of a linear MIMO system are real or in complex conjugate pairs.

Power series are one method [125] of defining functions and in particular for an arbitrary matrix L

$$\exp(tL) \triangleq \sum_{k=0}^{\infty} (tL)^k/k! \quad \text{with} \quad (tL)^o \triangleq I \qquad (2.4)$$

Like its "scalar relative," the above series is absolutely summable [112,125] for all time t, and therefore

$$\frac{d}{dt}[\exp(tL)] = \sum_{k=0}^{\infty} \frac{1}{k!}\frac{d}{dt}(tL)^k = L\exp(tL)$$

It follows that

$$\frac{d}{dt}\left[L^{-1}\exp(tL)\right] = L^{-1}\frac{d}{dt}[\exp(tL)] = \exp(tL)$$

or

$$L^{-1}\exp(tL) = \int \exp(tL)dt \qquad (2.5)$$

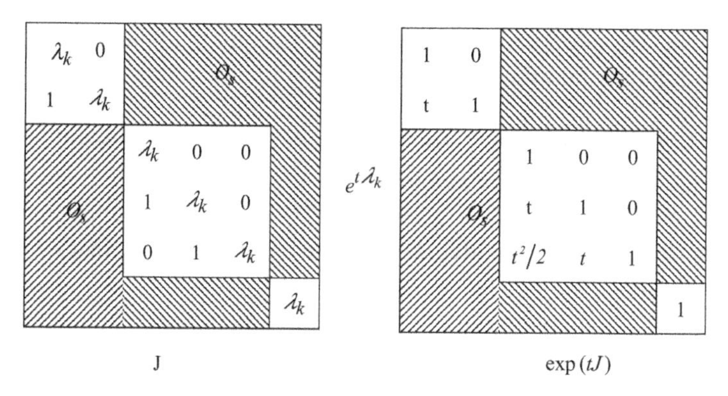

Figure 2.1 Jordan Sub-Blocks for the Eigenvalue λ_k

Application of the above to the Laplace transform of exp (tL) yields

$$\int_{o}^{\infty} \exp\left[-t(sI - L)\right]dt = (sI - L)^{-1}\exp\left[-t(sI - L)\right]_{\infty}^{o} = (sI - L)^{-1}$$

so that

$$\exp(tL) \text{ and } (sI - L)^{-1} \text{ are transform pairs} \tag{2.6}$$

For any matrix L it can be shown that a special linear change of Cartesian coordinates transforms L into the diagonal sub-block structure of its Jordan Form J [110,121]. Each block corresponds to a different eigenvalue and Figure 2.1 illustrates a typical sub-block of J and exp (tJ) for an eigenvalue λ_k. The literature [77,126] formulates various definitions of dynamic stability based on the mathematical concepts of a bounded variation about, or convergence to, a particular steady state. An appropriate criterion for present purposes is

$$\begin{gathered}\text{MIMO system (2.1) is stable if and only if for all } x_o \\ \text{and } u(t) \equiv 0 \text{ then} \\ \lim_{t\to\infty} x(t) = 0\end{gathered} \tag{2.7}$$

Because dynamic stability is patently independent of the choice of Cartesian coordinates, those creating a transition matrix in the Jordan

Form can be adopted. Accordingly, equations (2.2) and (2.6) with Figure 2.1 translate the above into

MIMO system (2.1) is stable if and only if all eigenvalues of its transition matrix have strictly negative real parts (2.8)

Equation (2.2) provides the output of a linear MIMO system explicitly as

$$\bar{y} = G(s)\bar{u} + C(sI - A)^{-1}x_o \quad \text{where} \quad G(s) \triangleq D + C(sI - A)^{-1}B \tag{2.9}$$

and $G(s)$ is termed the Transfer Function Matrix which in practice is usually square. By augmenting the state vectors, transfer function matrices $G_1(s)$ and $G_2(s)$ in series or parallel combine as those for SISO systems—except for commutivity. Specifically [122,134]

$$\text{In series}: G(s) = G_2(s)G_1(s) \text{ and in parallel } G(s) = G_1(s) + G_2(s) \tag{2.10}$$

A Resolvent $(sI - A)^{-1}$ for finite n-dimensions has the rational form [111]

$$(sI - A)^{-1} = \left(\sum_{k=1}^{n} s^{k-1}T_k\right) \Big/ \prod_{k=1}^{n}(s - \lambda_k) \quad \text{with} \quad \{T_{k\leq n}\} \text{ real matrices} \tag{2.11}$$

So the eigenvalues of state transition matrix are seen to be the poles of $G(s)$. Furthermore, by definition[3]

$\tilde{s}$ is a zero of $G(s)$ if and only if for some non-zero input $u(t) = \exp(\tilde{s}\,t)\,\tilde{u}$, no output response occurs, i.e., $y(t) \equiv 0$ (2.12)

[3] The classical SISO system notion that transfer function zeros are the roots of its scalar numerator polynomial is in fact equivalent to definition (2.12).

Though these zeros appear to pose an intractable calculation, they are in fact the roots of the zero polynomial [122,123,134]

$$Z(s) = \det(sI - A)\det G(s) \tag{2.13}$$

Thus problems of dynamic stability, etc., can be couched in the potent algebra of complex variables [113,114].

For instance, residue calculus implies that the pth output of a stable linear MIMO system for a solitary non-zero input component $u_k(t)$ approximates after a "long enough" time to

$$y_p(t) \simeq \text{Residue of } e^{st}G_{pk}(s)\bar{u}_k \text{ at poles of } \bar{u}_k \text{ only} \tag{2.14}$$

In particular for

$$u_k(t) = \exp(i\omega t); \quad \bar{u}_k = (s - i\omega)^{-1}; \quad \text{and} \quad u_q(t) \equiv 0 \quad \text{for } q \neq k$$

the pth component of the steady-state response is then evidently

$$y_p(t) \simeq G_{pk}(i\omega)e^{i\omega t} = |G_{pk}(i\omega)|\exp\left[i\omega t + iArgG_{pk}(i\omega)\right]$$

As real or complex pairs of eigenvalues are involved it follows from equations (2.9) and (2.11) that

$$G_{pk}(-i\omega) = |G_{pk}(i\omega)|\angle - ArgG_{pk}(i\omega)$$

Consequently, by virtue of system linearity, its steady-state response to

$$u_k(t) = \sin\omega t = \frac{1}{2i}[\exp(i\omega t) - \exp(-i\omega t)]; \quad \text{and} \quad u_q(t) \equiv 0 \quad \text{for } q \neq k$$

is

$$y_p(t) \simeq |G_{pk}(i\omega)|\sin\left[\omega t + ArgG_{pk}(i\omega)\right] \tag{2.15}$$

Naturally enough $G(i\omega)$ is termed the Real Frequency Response of a linear model, and it plays a pivotal role in questions of system stability.

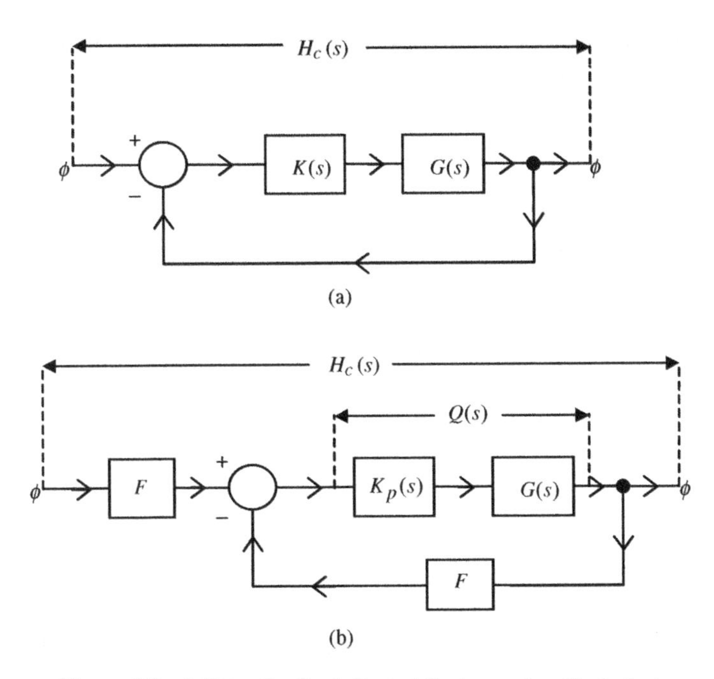

Figure 2.2 A Unity Feedback Control System and an Equivalent

For design purposes, the controller $K(s)$ in Figure 2.2a is partitioned into two parts

$$K(s) = K_p(s)F \quad \text{with } F \text{ a diagonal scalar gain matrix} \tag{2.16}$$

and the feedback system is very often reconfigured [123,134] as in Figure 2.2b. Engineering experience usually matches the number of control inputs to the outputs requiring control, so $K_p(s)$, F and $G(s)$ are usually square $(n \times n)$. The overall closed-loop transfer function matrix $H_c(s)$ is

$$H_c(s) = [I + Q(s)F]^{-1}Q(s)F \quad \text{where} \quad Q(s) \triangleq G(s)K_p(s)$$

but Figure 2.2b shows that stability etc., is determined by

$$H(s) = [I + Q(s)F]^{-1}Q(s) \tag{2.17}$$

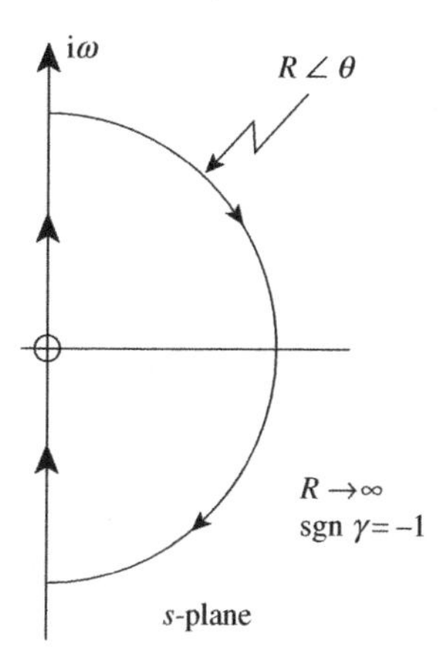

Figure 2.3 The Contour γ Enclosing Unstable Poles

However, for actual MIMO systems in particular, the inverse transfer function relationship [124,135,136]

$$\hat{H}(s) \triangleq H^{-1}(s) = F + \hat{Q}(s) \quad \text{with} \quad \hat{Q}(s) \triangleq Q^{-1}(s) \tag{2.18}$$

is evidently more tractable.

If $Q(s)$ for engineering purposes is close enough to diagonal, then design reduces to a number of independent SISO systems. For each SISO system its closed-loop poles and zeros are derived from equation (2.17) as the zeros of $1 + Q(s)F$ and $Q(s)$ respectively. The contour γ in Figure 2.3 encloses all the unstable closed-loop poles, and using the classical Encirclement Theorem [112–114] Nyquist in 1932 showed that

A Unity Feedback SISO system is stable if and only if the mapped contour $1 + Q(\gamma)F$ encircles the origin P_o times anticlockwise, where P_o is the number of unstable open loop poles.

A simple vector diagram shows that origin encirclements by$1 + Q(\gamma)F$ equate to encirclements of the critical point $(-1, 0)$ by $Q(\gamma)F$. Also in practice $Q(s) \to 0$ as $|s| \to \infty$, so the above reduces to

A Unity Feedback SISO system is stable if and only if its Real Frequency Response locus $Q(i\omega)F$ encircles $(-1, 0)$ anticlockwise P_o times (2.19)

Though some SISO defense systems have open loop poles at the origin due to kinematic integrations, these can be considered infinitesimally inside the stable region for design purposes.[4] Open loop systems are very often stable, and then the conformal mapping theorem [113,114,129] quantitatively relates transient closed-loop damping to the proximity of a $Q(i\omega)F$ locus to $(-1,0)$ in terms of Gain and Phase Margins [130,131]. Empirical rules [79,124] then also enable the design of analogue or digital controllers to achieve satisfactory closed-loop transient damping.

Presently MIMO feedback control systems can be designed in the frequency domain only if the pairs of transfer function matrices $[Q(s),H(s)]$ or $[\hat{Q}(s),\hat{H}(s)]$ are diagonally dominant [123,134]. Specifically, the magnitudes of their diagonal elements must strictly exceed over the *entire* γ-contour the sum of all others in the corresponding row or column. Diagonal dominance of $[\hat{H}(s),\hat{Q}(s)]$ can often be contrived and then confirmed by the superposition [82,157] of Gershgorin Discs on all $n \times n$ elements of an Inverse Nyquist Array $\hat{Q}(i\omega)$ [123,134]. The apparently simplistic replacement of rows or columns by linear combinations with others appears to be an effective first step. Other techniques for achieving diagonal dominance are fully described in the literature [123,134]. In essence these procedures correspond to a scalar matrix operating on the control error vector and this matrix is then incorporated in $K_p(s)$. Denoting diagonal elements of F and $\hat{Q}$ in equation (2.18) by f_k and $\hat{q}_{kk}(s)$, respectively, Rosenbrock's stability criterion for open loop stable dynamics is

A diagonally dominant Unity Feedback MIMO system is stable if and only if for all $k \le n$ *the origin encirclements by the locus* $\hat{q}_{kk}(i\omega)$ *equate to similarly orientated encirclements of* $(-f_k, 0)$ (2.20)

Thus diagonally dominant MIMO control systems can be engineered by well-established SISO techniques with readily computed Ostrowski Discs to assess residual loop interactions [82,134].

[4] Because compensating networks should move them further still to the left.

In general linear partial differential equations involve linear mappings between infinite dimensional vector spaces,[5] and their matrices do not always exist [111]. On the other hand, a finite number of linear ordinary differential equations correspond to linear mappings between finite dimensional spaces whose matrices always exist [110,111] and are computationally tractable. Heat diffusion in power plant metalwork is characterized by linear partial differential equations [224] and so might appear outside the prescribed framework of equation (2.1). However, the vast majority of energy transfer is associated with the smallest eigenvalue [117,118], so that ordinary differential equations become reasonable approximations for suitably sized segments of an Eulerian mesh [117]. Though partial differential equations represent steam generator and reactor dynamics, finite difference equations as a finite number of non-linear ordinary differential equations satisfactorily match practical tests [117]. Their linearization therefore accords with equation (2.1) and with Aizerman's conjecture, intact plant control via transfer function matrices can be engineered.

Practical studies [133,138–140] indicate that contriving diagonal dominance can require considerable skill: even with transfer function matrices much smaller than those for a complete power plant. Also, when manual intervention becomes imperative in some accidents, the control of an intact plant variable using a linear combination of several inputs appears to be humanly intractable. Moreover personal experience, exemplified by Section 3.4, is that ad hoc accident control cannot be conceived without an essentially one-to-one relationship between controlling and controlled plant variables. However, engineered rate constraints on nuclear plant temperatures etc. to ensure economic longevities or the intervention of trip circuits [127] intrinsically impose such essentially one-to-one relationships. As a result, though fossil and nuclear power plants are MIMO systems, their control can generally be addressed in terms of SISO theory as illustrated in Sections 3.2 and 3.3.

Theoretical concepts necessary to appreciate Chapters 2 and 3 have now been briefly outlined. Their first application is in the control of nuclear reactor power.

[5] A linear partial differential equation over all $t > 0$ has a countably infinite set of eigenvalues, whose eigenvectors are linearly independent [110–111].

2.2 MATHEMATICAL DESCRIPTIONS OF A NEUTRON POPULATION

Transport theory [58] offers the most accurate description of a reactor's neutron population in terms of a vector flux, but it has stringent computational demands. However, other than very close to strong absorbers or emitters,[6] neutronic velocity vectors are approximately isotropic and neutron migration can be readily computed when treated like the diffusion of gas molecules. Accordingly, with appropriate boundary conditions neutron conservation is characterized by a scalar neutron flux ϕ as [58]

$$D\nabla^2\phi + \sum\nolimits_a \phi + S = \frac{1}{V}\frac{\partial \phi}{\partial t} \tag{2.21}$$

where

ϕ – Scalar neutron flux $\triangleq$ Number of neutrons per square centimeter per second
D – Diffusion coefficient
$\sum_a$ – Macroscopic absorption coefficient
S – Expected neutron production rate per unit volume
V – Neutron speed in each chosen energy band of a simulation

Two- or three-dimensional multigroup[7] diffusion calculations of proposed core geometries have been validated by experimental zero-energy assemblies, and they have been proven successful in the United Kingdom for designing AGRs, SGHWR, PFR, and naval PWRs.

The fixed compact core geometries associated with fast reactors and PWRs have normalized neutron flux profiles that are largely governed by the escape of neutrons from the fissile core region, and so are substantially independent of output power. In addition fuel enrichment is deliberately increased toward the core periphery to "flatten" the radial flux profile and thereby enhance economics. These considerations intuitively suggest that the dynamics of these reactor types can be

[6] Control rods or a start-up source, for example.

[7] Neutronic energies quantized to replicate their slowing down, etc.

Table 2.1
Delayed Neutron Data for BWRs and PWRs with Uranium Fuel

Precursor group	1	2	3	4	5
Fraction (β_j)	0.00084	0.0024	0.0021	0.0017	0.00026
Decay constant (τ_j) (s)	0.62	2.19	6.50	31.7	80.2

closely approximated by one-dimensional distributed models [117], and experiments confirm this conjecture. Moreover the point kinetics model in Section 2.3 can also be derived [117] more rigorously by applying the analytical technique of adjoint (conjugate) linear mappings to these distributed model equations. This further simplification to a point model has proved sufficient for many control and overall plant simulations. However, because steam is a far weaker absorber than its liquid phase, the neutron flux profile in direct cycle systems (e.g., BWRs) changes materially with output power, so these reactor dynamics necessarily require the simultaneous solution of the distributed neutron diffusion and thermal-hydraulic equations [145]. Most neutrons (so-called prompt) are released at fission but a very small minority appear somewhat later as various fission products undergo radioactive decay. Table 2.1 lists the pertinent parameters for these delayed neutrons and their precursors. Later in Section 2.5 they are shown to influence reactor dynamics seemingly out of all proportion to their relative concentrations.

2.3 A POINT MODEL OF REACTOR KINETICS

Figure 2.4 depicts a conceptual model of nuclear reactor kinetics whose variables are defined by

> N—total number of free[8] neutrons in the reactor at any time t
>
> λ—life expectancy of a free neutron in the reactor
>
> $N + \delta N$—total number of free neutrons in the reactor at $t + \lambda$
>
> G—expected number of neutrons born after a fission

[8] Not temporarily withheld in precursors.

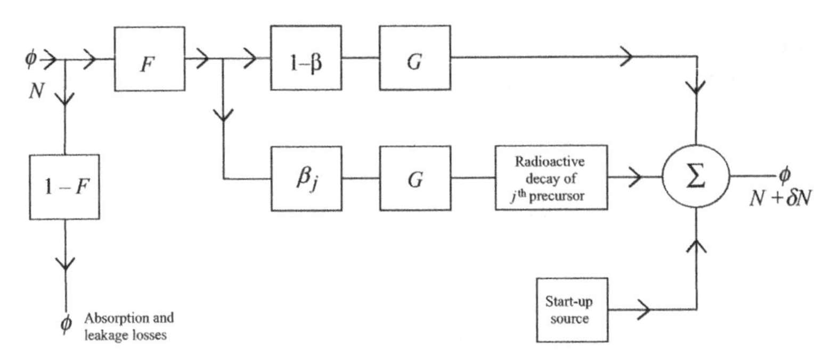

Figure 2.4 Conceptual Model of Neutron Kinetics

$1-\beta$—conditional probability of prompt fission with $\beta = \sum_{j=1}^{J} \beta_j$

β_j—conditional probability of creating a neutron precursor of the jth group

τ_j—time constant for the radioactive decay of the jth-precursor group

C_j—total number of neutrons temporarily withheld in the jth-precursor group

C_A—neutron production rate from the artificial start-up source

$1-F$—probability that a neutron is parasitically absorbed or escapes from the core

J—number of identifiably different precursor-groups

For clarity, just one group of delayed neutron precursors is shown in the diagram, though in practice a full description would generally involve no more than six.[9] As shown in Table 2.2, the above parameters depend on the type of reactor (fast or thermal), and also on its current geometry and temperature distribution. A 1-D diffusion model for tracking these parameters with changes of internal temperature distribution and control rod position is discussed in Ref. [117]. Because the expected lifetime λ of a free neutron is so much shorter than the time constants of the radioactive decay of precursors, and because these decay processes are Poissonian, then the number of precursor-atoms in the jth-group which release their

[9] Though Table 2.2 lists just five, a sixth results from the interaction of γ-rays with heavy water in an SGHWR or CANDU reactor.

Table 2.2
Some Pertinent Parameters of a Point Reactor Kinetics Model

	Thermal Reactor	Fast Reactor
$\sum_1^5 \beta_j \tau_j$ (s)	0.091	0.040
λ (s)	1.2×10^{-3}	3.0×10^{-7}
β	7.0×10^{-3}	3.5×10^{-3}
G for U-235	2.44	2.60
G for Pu-239	2.87	3.08

neutrons over the period λ is $C_j\lambda/\tau_j$. The number of these precursor-atoms created in the same interval is seen from Figure 2.4 to be $F\beta_j GN$, so that to a first approximation

$$\frac{dC_j}{dt} = F\beta_j GN/\lambda - C_j/\tau_j \quad \text{for} \quad 1 \leq j \leq J \tag{2.22}$$

Neutrons in a reactor core originate from prompt fission, the radioactive decay of precursors and the artificial start-up source. Reference again to Figure 2.4 enables the neutron population after the time internal λ to be derived as

$$N + \delta N = F(1-\beta)GN + \sum_{j=1}^{J} C_j\lambda/\tau_j + C_A\lambda \tag{2.23}$$

Because a target nucleus absorbs a neutron prior to its fission, the actual increase in the neutron population over this time interval is δN, so that to a first approximation[10]

$$\frac{dN}{dt} = [F(1-\beta)G - 1]N/\lambda + \sum_{j=1}^{J} C_j/\tau_j + C_A \tag{2.24}$$

Defining

$$K \triangleq FG \tag{2.25}$$

[10] Implicit in the notation of equation (2.22), whose left-hand side could otherwise be just δN.

reduces equations (2.22) and (2.24) to

$$\left.\begin{aligned}\lambda\frac{dC_j}{dt} &= K\beta_j N - C_j\lambda/\tau_j \quad \text{for} \quad 1 \leq j \leq J \\ \lambda\frac{dN}{dt} &= [K(1-\beta)-1]N + \sum_{j=1}^{J} C_j\lambda/\tau_j + \lambda C_A\end{aligned}\right\} \tag{2.26}$$

Some pertinent parameters for typical thermal and fast reactors are specified in Table 2.2, and the leakage and absorption factor $1 - F$ is derived from diffusion equation simulations as about 0.2. The much shorter life expectancy λ of a neutron in a fast reactor results from its much greater fuel enrichment of around 20% and the absence of a moderator.

It is seen from Figure 2.4 that

FGN = Number of prompt and delayed neutrons created during the lifetime of their free parents

so from equation (2.25)

$$K = \frac{\text{Number of prompt and delayed neutron births}}{\text{Number of their free parents}} \tag{2.27}$$

In that equation (2.25) allows for both the escape (leakage) and parasitic capture of neutrons, the above equation evidently justifies the description of K as the effective multiplication factor. If the free neutron-parents produce their equal number of prompt and delayed offspring so that the effective multiplication factor is unity, then the population appears stable. Indeed neglecting the relatively small contribution of the start-up source, equation (2.26) confirms that

$$K = 1 \tag{2.28}$$

is a necessary and sufficient condition for the neutron population in a reactor to remain numerically constant.

Because the combined mass of fragments from a fission is less than that of their fissile parent atom [58], the deficit δm appears in the form of their kinetic energy $\delta m\, c^2$. Collisions with surrounding materials rapidly degrade this into heat, which for either U-235 or Pu-239 amounts to about 32 pJ per fission. It follows therefore from Figure 2.4 that

$$\text{Reactor power} = 32\frac{FN}{\lambda} \times 10^{-12}\,\text{W} \tag{2.29}$$

Nuclear power reactors are seen, therefore, to have enormous neutron populations, and for simulation purposes equations (2.26) are conveniently scaled, say by trillions of neutrons.

2.4 TEMPERATURE AND OTHER OPERATIONAL FEEDBACK EFFECTS

With material power production, temperatures and other phenomena change the effective multiplication constant K. Higher temperatures increase the vibrations of component nuclei and decrease their densities. One effect is to widen the effective resonance absorption bands of U-238. Because neutrons are slowed down by scattering in energy steps many times larger than these resonance widths, their non-fissile capture rates by U-238 increase with fuel temperature [58]. This negative reactivity feedback mechanism is called the Doppler Effect,[11] and it is an important re-stabilizing influence on the neutron populations of both fast and thermal reactors. Indeed fast reactor designs in particular have progressively evolved with greater U-238 content and less energetic neutrons so as to exploit the effect. Denoting the mean absolute temperature of the fuel by T, then in terms of the effective multiplication factor it is found that under normal operation

$$-T\frac{dK}{dT} = d \quad \text{with} \quad d > 0 \tag{2.30}$$

where the Doppler Constant d is specific to a reactor design. Equation (2.30) shows that due to the Doppler Effect the effective multiplication factor is inversely proportional to the logarithm of the mean fuel-

[11] Doppler broadening alone affected restabilization of a growing neutron population in the University Argonaut "zero power" reactor at Risley.

temperature ratio. Other temperature feedback effects on reactor dynamics are generally associated with

- variations in coolant-density either by thermal expansion or by vaporization that alter the absorption or moderation of neutrons
- thermal expansion of the fuel and control rods similarly alters their macroscopic cross-sections
- thermal expansion of a moderator leading to faster neutrons (i.e., a "harder spectrum") with more resonance absorptions by U-238 nuclei
- thermal expansion of the fuel cladding (zircalloy or stainless steel tubing) whose "bowing" excludes coolant.

These interactions and their associated time delays are significant features of nuclear reactor dynamics.

Neutron absorption is also significantly affected by the in-pile dwell time of the fuel, and its preceding 7 to 47 h power history due to developing concentrations of Xe-135 and Sm-149. Glasstone and Edlund [58] quantify the former as the more dominant "neutron poison," and it is the daughter of the fission product I-135 whose half-life is 9.17 h [76]. Whilst a reactor is at power, Xe-135 is transmuted ("burned up") faster by neutron capture than by its natural decay rate. However, in the event of a complete operational shutdown (trip), its concentration increases progressively even for as long as 12 h because of the relatively faster disintegration of I-135. As a result, it could be impossible to restart power production in this period unless sufficient latent reactivity has been held in reserve (i.e., by pre-trip control rod insertions).

The direct cycle RMBK reactor at Chernobyl was moderated by both graphite and a light water coolant, which was partially converted to steam for electric power production. Unlike heavy water, light water is both an effective moderator and absorber,[12] so its conversion to less dense steam reduces both neutron moderation and absorption. In the unauthorized fateful incident, the night-shift operators cancelled trip settings and withdrew all 211 control rods [12]. Progressive reductions in inlet-water flow then resulted in a growing volume of steam in the reactor core, and a net reduction in neutron absorptions eventually

[12] See Section 1.8.

occurred because moderation by its graphite was still sufficient. In this way a progressive positive feedback process was initiated that produced an exponentially increasing reactor power with the prompt time constant [58,80,117] of around 1 ms, and some hundred times [12] full-rated power resulted. Subsequently by

i. the contact of molten fuel with liquid water [59,212],
ii. the accretion of then more mobile fission products into sizeable bubbles, whose external pressures increase as the surface tension effect is reduced [210], and
iii. hydrogen production [12] from the chemical reduction of steam by graphite,

there was the recorded explosive destruction of the site. A nuclear explosion was *not* involved.

Due to astute design of core-lattice geometry [61] and fuel enrichment, light water reactors outside Russia have always been designed to become under-moderated with increasing steam production in order to stifle such potentially explosive events. Indeed a negative power-reactivity coefficient is a necessary prerequisite for licensing by European Regulatory Authorities.

2.5 REACTOR CONTROL, ITS STABLE PERIOD, AND RE-EQUILIBRIUM

Section 2.4 describes how power production changes the effective multiplication factor of a nuclear reactor so that

$$K = K(N) \tag{2.31}$$

As a result equation (2.26) is non-linear. However, for control analysis it is sufficient to examine this equation for small perturbations about an operating point for which the effective multiplication factor is a constant derived from neutron diffusion and possibly thermal-hydraulic calculations. Effecting the Laplace transformation of equation (2.26) and involving the definition

$$\beta = \sum_{j=1}^{J} \beta_j \tag{2.32}$$

yields

$$\left[\lambda s - (K-1) + K\sum_{j=1}^{J}\left(\frac{\beta_j \tau_j s}{\tau_j s + 1}\right)\right]\bar{N} = \lambda C_A/s \tag{2.33}$$

For analytical purposes, equation (2.33) is more conveniently expressed in terms of the core reactivity

$$\rho \triangleq (K-1)/K \tag{2.34}$$

to give

$$\left[\lambda s + \left(\frac{\rho}{\rho - 1}\right) + \left(\frac{1}{1-\rho}\right)\sum_{j=1}^{J}\frac{\beta_j \tau_j s}{\tau_j s + 1}\right]\bar{N} = \lambda C_A/s \tag{2.35}$$

The algebraic roots of

$$F(s) \triangleq \lambda s - \rho(\lambda s + 1) + \sum_{j=1}^{J}\frac{\beta_j \tau_j s}{\tau_j s + 1} = 0 \tag{2.36}$$

define the poles of the neutron population's kinetics, and the real root closest to the origin is termed dominant whose reciprocal T^* defines the reactor period. As reactivity is increased from zero, the dominant pole moves into the right half s-plane causing the neutron population to diverge exponentially in the form $N_0 \exp(tT^*)$. The location of the dominant pole s^* can be determined iteratively using the Newton–Raphson algorithm

$$s^* = \tilde{s} - F(s) \Big/ \left(\left.\frac{dF}{ds}\right|_{s=\tilde{s}}\right) \tag{2.37}$$

where

$\tilde{s}$—an estimated location of the dominant pole and

$$\frac{dF}{ds} = \lambda(1-\rho) + \sum_{j=1}^{J}\frac{\beta_j \tau_j}{(\tau_j s + 1)^2} \tag{2.38}$$

For small enough reactivities, the reactor period is estimated from the above equations as

$$T^* = \left[\lambda(1-\rho) + \sum_{j=1}^{J} \beta_j \tau_j\right] \Big/ \rho \tag{2.39}$$

which from Table 2.2 further approximates to

$$T^* = \left[\sum_{j=1}^{J} \beta_j \tau_j\right] \Big/ \rho \tag{2.40}$$

so that the radioactive decay periods of the precursor groups govern the growth of a neutron population. On the other hand for large reactivities the dominant root of equation (2.36) is far removed from the $\{1/\tau_j\}$, and under these conditions it behaves as

$$\lambda s - \rho(\lambda s + 1) + \beta = 0 \tag{2.41}$$

The corresponding reactor period is then asymptotic to

$$T^* \approx \lambda(1-\rho)/(\rho - \beta) \tag{2.42}$$

which is dominated by the lifetime of prompt neutrons, because as seen from Figure 2.4 these alone constitute a self-sustaining subpopulation. Typical data in Table 2.2 and the above analysis enable a simple digital computation of the stable reactor period as a function of normalized reactivity (ρ/β). Reactivity are often quantified in this normalized form for operational purposes, and the so-called prompt critical value of unity is ascribed a magnitude of one dollar ($1) with corresponding subdivisions of cents. Once a reactor enters the super-prompt critical regime, the reactor period is seen from Figure 2.5 to decrease dramatically: especially for fast reactors.[13] Bearing in mind the typical rate

[13] Fuel enrichments in thermal and fast reactors are some 3 and 20%, respectively. The inadvertent insertion of a fuel rather than a U-238 breeder pin in a fast reactor could cause a potentially serious neutron excursion, so extreme care is necessary in their refuelling.

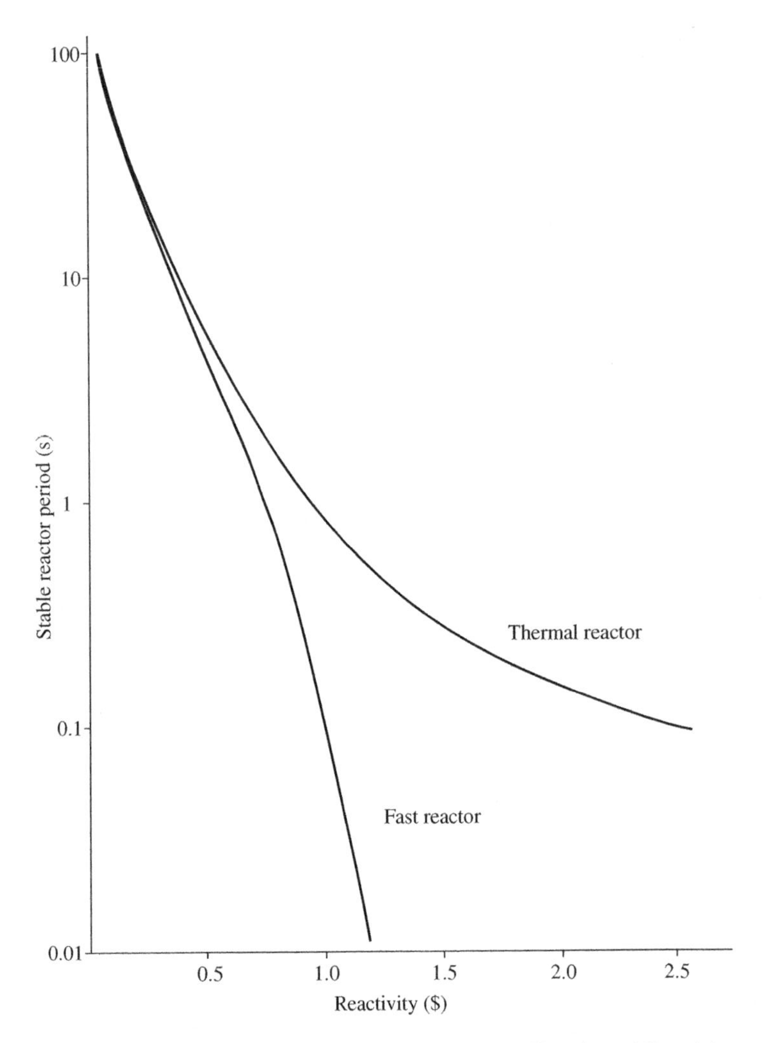

Figure 2.5 Typical Variations of Reactor Period as a Function of Reactivity

constraints associated with induced thermal stresses in a power plant, and the interval necessary for emergency intervention, then reactivity changes in normal operation must be restricted to a few cents (1/100th of 1$) to achieve a stable reactor period of no less than about 30 seconds [80,117]. After circumspect increases in reactor power by restricting withdrawal of control rods, the negative reactivity feedbacks described in Section 2.4 restablize a neutron population. In terms of equation (2.40) such increases in reactor power or neutron population correspond to an infinite reactor period T^* with $K = 1$ or $\rho = 0$. An alternative viewpoint from equation (2.36) is that a neutron population (reactor power) in equilibrium corresponds to the dominant pole at the origin with $\rho = 0$.

CHAPTER 3

Some Power Station and Grid Control Problems

3.1 STEAM DRUM WATER-LEVEL CONTROL

Grid-connected power stations are operated in either decoupled or coupled mode [80,117], which describes the form of their response to grid frequency variations[1] from changes in network demand. These alternatives are illustrated in Figures 3.1 and 3.2 where $C_1(s)$ and $C_2(s)$ represent controller transfer functions. By creating strict limits on steam plant temperature and pressure variations, decoupled control promotes steam turbine efficiency and boiler longevity (e.g., by reducing the exfoliation of tube magnetite deposits). Consequently, decoupled control is generally the economic choice for large base-load stations having a high thermal efficiency and capital investment. With a coupled control scheme, a fossil or nuclear heat source is modulated to sustain boiler pressure when changes in Grid frequency operate the turbine control valve. Because Grid frequency fluctuations widely outpace heat-source dynamics, boiler-pressure changes can be offset only by exploiting the thermal energy stored in the rest of the plant. Such spontaneous changes of stored energy in some plant components are readily accommodated and their responses help to relieve thermally induced stresses in other less robust items [141].

[1] See Section 3.3.

Nuclear Electric Power: Safety, Operation, and Control Aspects, First Edition.
J. Brian Knowles.
© 2014 John Wiley & Sons, Inc. Published 2014 by John Wiley & Sons, Inc.

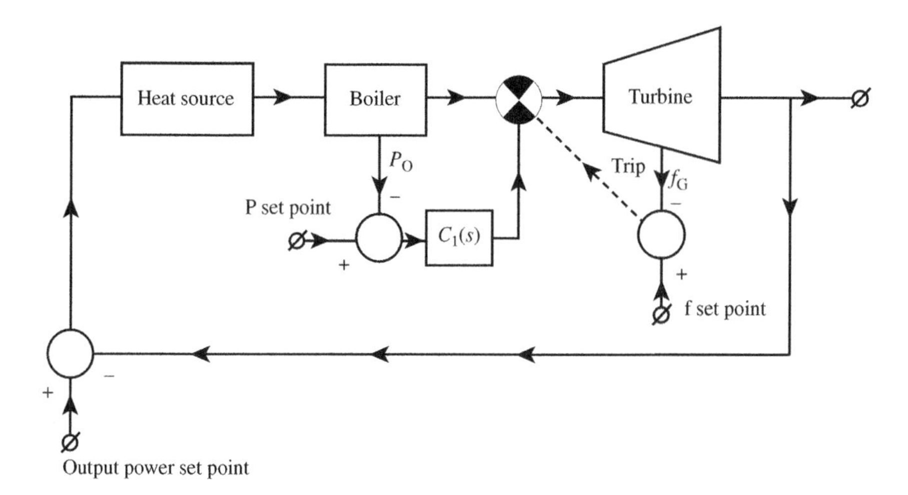

Figure 3.1 The Basic Decoupled Control Scheme

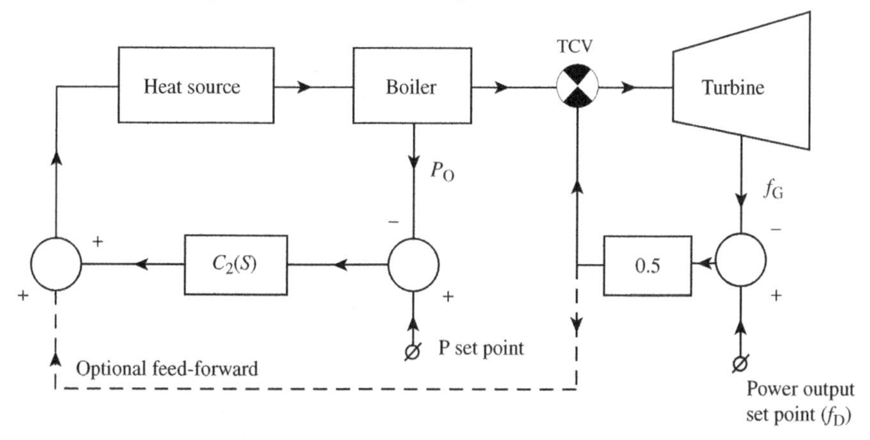

Figure 3.2 The Coupled Control Scheme

The evaporator section of the La Mont boiler system [117] in Figure 3.3 produces a low steam quality ($\lesssim 10\%$) flow into a large steam drum which separates the saturated steam for superheating. Feed water to match steam generation is usually injected downwards from a sparge pipe near the base of a drum. A Richardson number analysis [117] establishes that its mixing with less dense saturated liquid from the evaporator is thereby largely suppressed,[2] so the drum's liquid content remains markedly stratified.

[2] Similar situations occur in domestic hot water cisterns and in salt fingers around river estuaries.

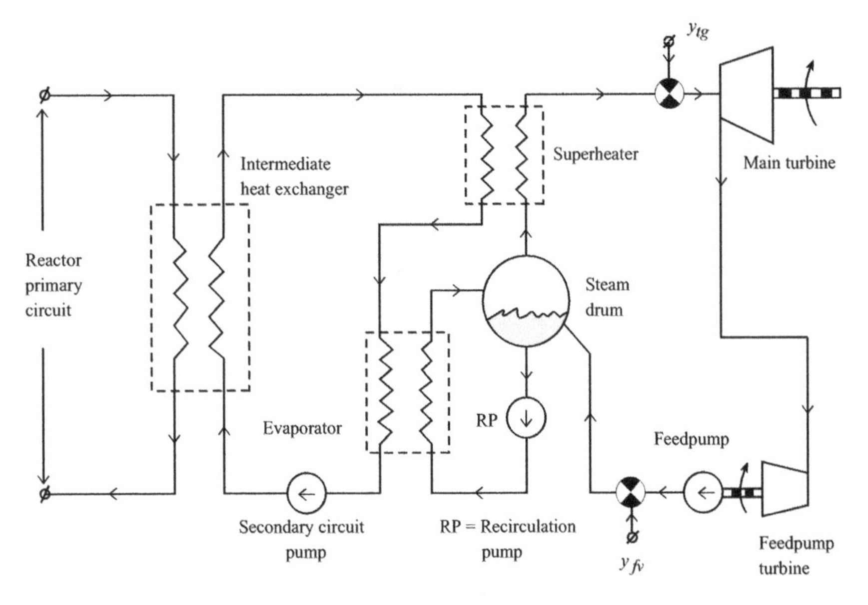

Figure 3.3 Schematic Drum-Level Control Problem

When the turbine control valve opens under coupled control, the increased mass flow rate of steam to the turbine reduces boiler pressure, so some of the upper layer of saturated water in the drum flashes rapidly into steam to preserve thermodynamic equilibrium. This extra steam partially supports system pressure, and there is a corresponding fall in drum water level. Because a large enough reduction would cause the drawdown of steam into the recirculation pump [117], drum water-level control is necessary to prevent cavitation damage to its impeller. On the other hand, too high a water level would impair steam separation leading to damaging thermal shocks to the superheater tubing.

Figure 3.3 depicts the drum water-level control scheme for a 250 MW(e) fast reactor nuclear power plant. The intermediate heat exchangers (IHXs) provide an additional safety barrier to obstruct an explosive ingress of water into the sodium-cooled reactor circuit. Because improvements in station efficiency of as little as 0.1% are financially material, the feed pump was driven from the same high-pressure steam supply as the main turbine to exploit this opportunity. Under coupled control a reduction in Grid frequency opens the turbine control valve via y_{tg} and the drum pressure falls as a result of the increased steam flow. Its saturated surface-water then flashes very rapidly into steam, so a correspondingly fast increase in feed-water flow is necessary to prevent the drawdown of steam. However, because the driving pressure to the feedpump turbine has reduced, the feed-flow is reduced just when it's needed. Hence, the main turbine and feedpump turbine control-valve settings y_{tg} and y_{fv} are strongly interactive making the plant's transfer function matrix far from diagonally dominant. It can be readily appreciated therefore that SISO control system design techniques proved unsatisfactory. By consummate skill, Hughes [81] devised a broadly satisfactory MIMO control system design using a 3×3 matrix controller with proportional plus integral diagonal elements. Nevertheless the incipient deployment of an independent electrically driven feedpump would have circumvented the problem to allow largely independent SISO control schemes and a more transparent design of ad hoc accident management strategies. This example highlights the importance of engineering insight and awareness of plant operating conditions: thereby demonstrating the industry specific nature of control engineering. Finally, it should be noted that the relatively high initial capital and low fuel costs of nuclear stations usually favours their top merit order placings with operation at maximum power under decoupled control.

3.2 FLOW STABILITY IN PARALLEL BOILING CHANNELS

Flow instability in the boiling channels of fossil or nuclear plants would soon lead to boiler tube or fuel pin ruptures from the thermally induced stresses. Experiments show that channel flow oscillations with a period of 1 to 10 s can exist under constant inlet and outlet pressures. A resolution of this paradox is obtained by considering the transport delays and changes of thermodynamic phase along a boiling channel [80,117]. Qualitatively, an inlet mass flow perturbation $\delta W_1 \sin \omega t$ in Figure 3.4 persists over the largely incompressible liquid-phase region to produce a simultaneous differential pressure drop of $\delta P_L \sin \omega t$. However, due to its compressibility, the average acceleration of the two-phase region and its corresponding pressure perturbation $\delta P_{2\phi}$ suffer a significant delay. A similar argument applies a fortiori to the steam region, whose average acceleration and differential pressure change δP_s are still further delayed. The phasor diagram in Figure 3.4 reveals qualitatively that a flow oscillation can exist with no change in

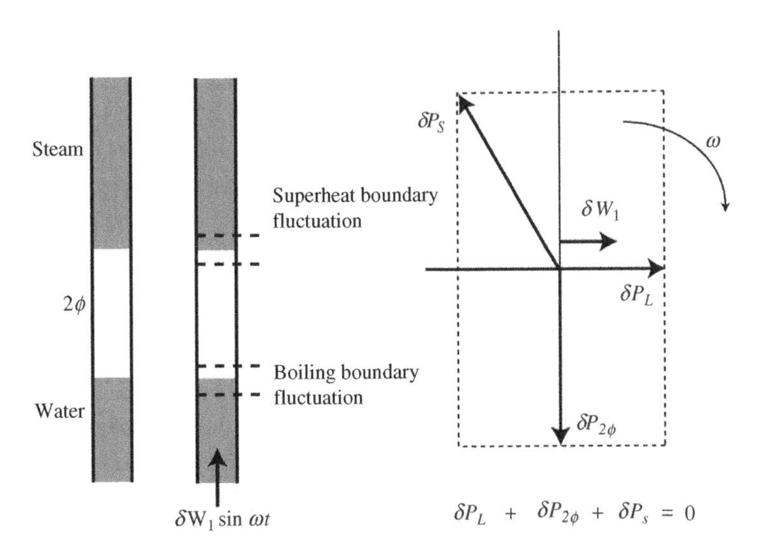

Figure 3.4 Relating to Parallel Channel Flow Stability

the differential pressure across a boiling channel. Flow stabilization can then only be achieved by inserting inlet ferrules (gags), which increase the liquid-phase component of a differential pressure change. As can be inferred from Figure 3.4, this artifice pulls the Nyquist diagram away from the critical (−1,0) point. Though feedpump power is relatively small,[3] quite small changes (e.g., 0.1%) in station efficiency are material [80], so the ferrules must be designed to be sufficient for purpose and little more.[4] Engineering experience and comprehensive non-linear simulations are now shown to simplify a quantitative analysis of the problem.

In response to an inlet-flow perturbation vector $\delta W_1(t)$ in Figure 3.5 define

$$u = \begin{pmatrix} \text{Primary inlet mass flow perturbation} \\ \text{Primary inlet temperature perturbation} \\ \text{Water-side inlet mass flow perturbation} \\ \text{Water-side inlet temperature perturbation} \\ \text{Water-side inlet pressure perturbation} \end{pmatrix} \tag{3.1}$$

[3] Some 2 MW in a 1000 MW(e) station.

[4] Allowances for in-service erosion and tube-to-tube variations.

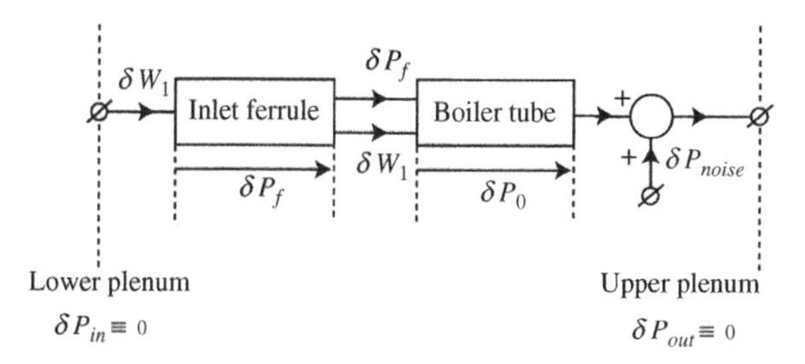

Figure 3.5 Parallel Channel Stability Model

and the incremental outlet flow vector of the boiling channel by

$$y = \begin{pmatrix} \text{Primary outlet mass flow perturbation} \\ \text{Primary outlet temperature perturbation} \\ \text{Water-side outlet mass flow perturbation} \\ \text{Water-side outlet temperature perturbation} \\ \text{Water-side outlet pressure perturbation} \end{pmatrix} \tag{3.2}$$

which are related by a matrix transfer function $T(s)$. During the observed period of flow oscillations, water-inlet temperature is effectively buffered by a large mass stored in the feed train [141], and inlet pressure by the feedpump circuit. Also at high subcritical pressures, heat transfer and thermodynamic states along a channel are essentially unaffected [64] by the relatively small pressure changes, and those across a ferrule simply add to the overall differential pressure perturbation [117]. Both inlet and outlet primary-side variables are maintained effectively constant by virtue of their: thermal capacities, pump speed, and reactor reactivity settings.[5] Accordingly the perturbed inlet mass flow vector to a channel reduces to

$$u = (\text{Water-side inlet mass flow perturbation}) \tag{3.3}$$

Thus the outlet water-side pressure perturbation δP_0 is given for practical purposes by the SISO transfer function relationship

$$\delta \bar{P}_0 = T_{53}(s)\delta \bar{W}_1 \tag{3.4}$$

[5] This conclusion applies to both fossil and nuclear plants.

With large amplitude mass flows, the pressure drop across a ferrule is [304]

$$P_f = P_f\left(W_1^2/\rho_L; \text{Ferrule Geometry}\right) \tag{3.5}$$

but because the density of liquid-phase water is essentially constant under all normal operating conditions, incremental pressure and mass flow changes for a ferrule are related by

$$\delta P_f = -K_f \delta W_1; \quad K_f > 0 \tag{3.6}$$

where K_f is a constant specific to a ferrule's geometry and the inlet flow W_1 at the selected output power. It follows from Figure 3.5 that no change in overall differential pressure occurs when

$$0 \equiv \delta \bar{P}_{\text{out}} = -K_f \delta \bar{W}_1 + T_{53}(s)\delta \bar{W}_1 + \delta \bar{P}_{\text{noise}} \tag{3.7}$$

where the noise term δP_{noise} arises from fluid turbulence and variations in pump speed induced by Grid-frequency fluctuations. Incremental flow stability by a choice of ferrule geometry can therefore be engineered from

$$\delta \bar{W}_1 = \left[\frac{1}{K_f + (-T_{53}(s))}\right] \delta \bar{P}_{\text{noise}} \tag{3.8}$$

By analogy with a SISO unity feedback system, Davis and Potter [83] originally in the context of SGHWR described $-T_{53}(s)$ as the "open loop" transfer function, and by analytical linearization derived transfer functions for the three different water-phase zones in Figure 3.4. Suitable ferrule geometries to cope with different load factors and flow distribution in the lower plenum were then selected by Nyquist diagram techniques. As an alternative, Knowles [117,128] perturbs comprehensive non-linear simulations to achieve the same end: but on the basis of the above more rigorous state vector formulation. Nevertheless, the original simplified analysis is sound and it illustrates the considerable simplification achievable by engineering insight and industry specific experience. However, as with all linearized models, confirmation by a full non-linear simulation is absolutely necessary.

3.3 GRID POWER SYSTEMS AND FREQUENCY CONTROL

Grid-connected power stations form a diverse interconnection of fossil, nuclear and renewable units whose objective is to meet the area's power demands as safely, economically and securely as feasible. On a continuous basis centralized management selects generation from those available units best able to meet these objectives. Sinusoidal ac power at nominal frequencies of 50 or 60 Hz has many advantages from the viewpoints of generation, transmission and utilization [35]. Specifically, two-pole cylindrical alternators with water-cooled conductors provide the highest commercially available power generation per unit volume, but even these machines are limited to around 3000 or 3600 rpm which corresponds to 50 or 60 Hz respectively. With present units of between 100 and 660 MW this rotational speed is near optimum for steam turbine efficiency and blade reliability. Consequently a turbine and an alternator can be directly coupled together with a bolted flange to avoid the complexities and inefficiencies of high-power gear trains. Electrical transmission losses countrywide are reduced by the use of high voltage (e.g., 400 kV) to current ratios, but use in industrial and domestic situations requires relatively lower voltages (440 V or 230 V). For frequencies of 50 or 60 Hz, transformers provide a highly efficient ($\gtrsim$96%) and reliable execution of this task.[6] However, as explained in the context of equation (1.62), high voltage dc is more cost-effective for cable transmission. A similar argument for dc transmission also applies to very long ($\sim$650 km) overhead lines for which corona losses [147] increase with line voltage[7] and length.

Figure 1.1 illustrates the partially predictable seasonal and daily changes in the power demands on a Grid network. In addition, there are unforeseen material fluctuations induced for example by the start-up or shutdown of large industrial plant, or the substantial loss of a 400 kV Supergrid transmission line. Consequently, instantaneous electricity generation and demand cannot be identically matched by pure

[6] Some aircraft systems use 400 Hz to reduce core size (i.e., weight), but hysteresis losses are correspondingly greater [35].

[7] Electric intensity is inversely proportional to conductor diameter, so each phase of an overhead supergrid line in the United Kingdom is a tightly bunched bundle of four conductors.

prediction, and thermal constraints also restrict each station's rate of change of power.[8] As electricity cannot be stored in the required quantities, a mismatch between instantaneous Grid generation and demand must be accommodated in the short term by thermal energy stored in the coupled generating units and in the rotational energies of all Grid-connected generators and motors. Because the synchronizing torque per degree electrical of an ac machine is so large compared to its inertia [35], all directly connected ac machines can be considered for present purposes to be "locked" together at a synchronous speed Ω rad/s given by

$$\text{Synchronous speed } (\Omega) = 2\pi \times \text{Grid frequency } (f_G)/\text{pole-pairs of a machine} \quad (3.9)$$

where f_G — Grid frequency (Hz).
Thus mismatches in Grid power appear throughout as a common frequency fluctuation about the nominal, and three principal reasons for its tight constraint now follow.

Firstly, each turbo-alternator is a multi-machine system of inertias linked by resilient shafts and therefore exhibit mechanical resonances at certain critical speeds [148] above and below the nominal 3000 or 3600 rpm. Unless the Grid frequency is controlled within narrow limits, these resonances could persist long enough to inflict serious damage. In practice during the Grid synchronization of a turbo-alternator these resonant speeds are accelerated through as quickly as possible.

Secondly, a large number of industrial and domestic consumers are still metered by electro-mechanical units which were installed by virtue of their good stability of calibration, wide measurement range and low cost of mass production [149]. Measurement accuracy with these single- and three-phase induction instruments depends on maintaining a 90° phase relationship between line voltage and the voltage coil's magnetic flux. Though a degree of compensation is provided by "shaded poles," measurement errors still occur when the frequency deviates from the calibration frequency of 50 or 60 Hz. Because the United Kingdom's national electricity consumption is currently of the order of 350 TWh/year, even very small errors are fiscally significant.

[8] Allowable variations in the axial temperature profile of the SGHWR steam turbine corresponded to between 2 and 3 MW per min.

Accordingly, UK consumers are protected by a Parliamentary Statute that requires the 24-h average deviation to be within ± 0.5 Hz, though National Grid plc self-imposes stricter limits of ± 0.2 Hz. Measurements [151] in 1972 characterized UK Grid fluctuations by a Normal distribution having a standard deviation of 0.05 Hz, which is consistent with the now continuously updated data on the Internet [150]. Safety trip limits for UK steam turbines impose operation between 48 and 52 Hz.

Finally, synchronous[9] or induction motors [35] are generally used to drive power station boiler feedpumps, whose pressure rise is approximately proportional to the square of their rotational speed. When power demand exceeds generation, the Grid frequency and feedpump speed fall, so boiler pressures are reduced contrary to the required increased steam flow and alternator output. On the other hand, when power demand is less than generation, feedpump speed rises so boiler pressures increase, and the life expectancy of turbine blading in its low-pressure cylinder could thereby be prejudiced by a potential over-expansion of the steam [117]. A suitably controlled boiler inlet-pressure is therefore necessary, so an excess feedpump pressure must be developed and the flow throttled to provide the required operating conditions.[10] Thermodynamics show that the steady-state power required for an incremental pump pressure change δP is

$$\text{Pumping power} = (W/\eta\rho)\delta P$$

where

W — mass flow rate (kg/s)

η — pump efficiency

ρ — water density (kg/m^3)

The above conditions imply that a +1% frequency deviation corresponds to an extra pumping-power loss of about 1.6 MW(e) for a 1200 MW(e) station having an ac motor-driven feedpump [80].

[9] Used as a means of Network power factor correction via their dc excitation [35].

[10] See Figure 3.9.

The required control of Grid frequency is achieved by closely balancing instantaneous generated power with demand. For this purpose previous statistics as well as meteorological forecasts and mass entertainment data are involved to continuously predict demand so as to accommodate intrinsic plant start-up delays and thermal rate constraints. In the United Kingdom, regional quotas are allocated on the basis of these predictions with consideration for plant outages and the overall security of supply. Accordingly, a Grid control region has more nominal capacity than historic demands. Individual stations are initially selected for operation by regional controllers in terms of a Merit Order based on fuel costs (£ per kWh) and reliability. This selection clearly favors relatively high capital but low fuel-cost nuclear stations, though these can no longer meet the minimum UK consumption. Thus contributions to the daily predicted load are required from fossil and renewable plants. When wind turbines are available, they too appear as an economic option for this purpose. However, in a UK winter, extensive areas of meteorological high pressure would render a large number of turbines impotent, so that fossil-fired units are required for balancing during this season of greatest demand. As described in Section 1.7 combined cycle gas turbine (CCGT) plants have progressively replaced less thermally efficient and more polluting end-of-life coal-fired stations since 1991. Though these factors in part favor the selected operation of CCGT units, the principally coal-fired 3960 MW(e) Drax plant is usually operational by virtue of its high capacity factor ($\simeq$75%) and relatively high thermal efficiency[11] ($\simeq$40%). Due to the absence of ponderous coal pulverizers and a more favorable fuel-combustion chemistry,[12] CCGT generation has the additional advantage of faster dynamics for meeting major predicted and unscheduled load changes. Consequently a number of CCGT stations are operated at around 75% of full-load (a spinning reserve) or at almost zero output (hot starts) to meet these load changes. Currently the United Kingdom has access to some 650 MW(e) of auxiliary diesel or gas turbine units along with 1800 MW(e) from the Dinorwic pumped storage scheme having a 10 s access time [40], and 2000 MW(e) of rapidly disconnectable load. With many individual stations maneuvering to effect an

[11] For a coal-fired plant.

[12] Coal combustion is initially endothermic.

instantaneous Grid power balance there is clearly the problem of overall network stability to be addressed.

In this context, the energy stored in an inertia (I) rotating at Ω rad/s is

$$E = {}^1\!/_2 I\,\Omega^2$$

so the rate of change of energy as it is delivered or withdrawn is

$$\text{Power} \triangleq \dot{E} = I\,\Omega\frac{d\Omega}{dt} \tag{3.10}$$

As described above, the speeds of all Grid-connected units can be considered locked together at the existing synchronous Grid frequency f_G. Because Grid frequency is necessarily maintained within ± 0.5 Hz about the nominal 50 or 60 Hz, equation (3.10) can be approximated by

$$\text{Grid power perturbation} \quad \delta\,\text{Power} = KR_T\dot{f}_G \tag{3.11}$$

where R_T is the sum of name-plate ratings for all synchronized generators and motors.[13] An allowance for some multi-pole pair units is accommodated by

$$0.2 \lesssim K \lesssim 0.4 \text{ per VA of } R_T \tag{3.12}$$

Coupled (or boiler-follows-turbine) stations buffer frequency fluctuations, but decoupled (or turbine-follows-boiler) stations contribute only to R_T. Figure 3.6 illustrates stability considerations for a hypothetically isolated coupled controlled station. As a result of meeting physical rate constraints and a circumspect engineering design,[14] its open loop frequency deviation to output-power transfer function can be approximated by a SISO function $H(s)$ which relates just the measured frequency deviation through the turbine control-valve dynamics to a release rate of stored energy in the plant. By means of comprehensive non-linear simulations, a $set\{H(iw)\}$ can be derived for a representative number of load factors including synchronization. Simulations then confirm that the isolated station can deliver demanded power up to its

[13] An electrical machine carries a nameplate stating its (nominal) safe continuous rating, e.g., 33 KV; 500 MVA.

[14] Refer to the fast reactor drum water-level control by Hughes in Section 3.1.

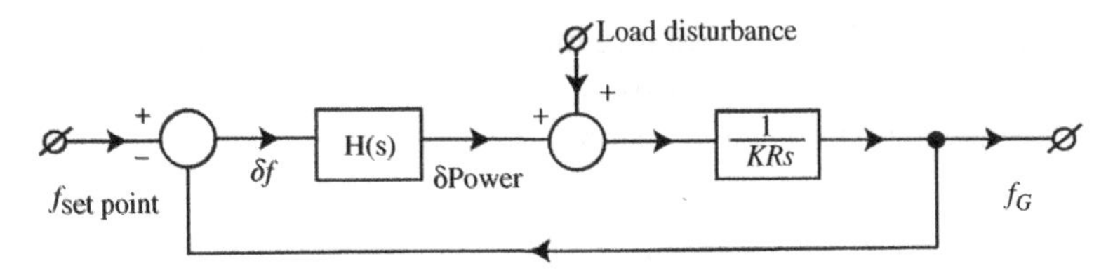

Figure 3.6 Frequency Control Stability of an Isolated Station

name plant rating, but the impact of its Grid connection on the stability of the overall network needs to be addressed as in Figure 3.7.

Intuitively it might seem that a parallel combination of individually stable isolated units would always be stable in the coupled-control mode, but theoretically this is untrue. Consider just two such stations with identical nameplate ratings R, but with different transfer functions $H_1(s)$ and $H_2(s)$. Equation (2.10) and Figure 3.7 yield the open loop Real Frequency Response of this hypothetical arrangement as

$$[H_1(i\omega) + H_2(i\omega)]/i\omega K2R = {}^1/_2[H_1(i\omega)/i\omega KR + H_2(i\omega)/i\omega KR]$$

If at some frequency $\hat{\omega}$ the individual station responses were complex conjugates of each other, then the open loop response function of their parallel combination would be

$$\text{Re}\ [H_1(i\hat{\omega}/i\hat{\omega}KR] \triangleq \text{Re}\ [H_2(i\hat{\omega})/i\hat{\omega}KR]$$

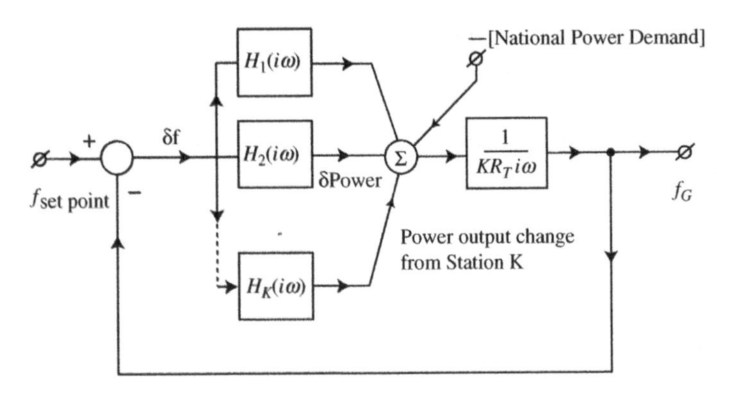

Figure 3.7 Grid Network Stability Model

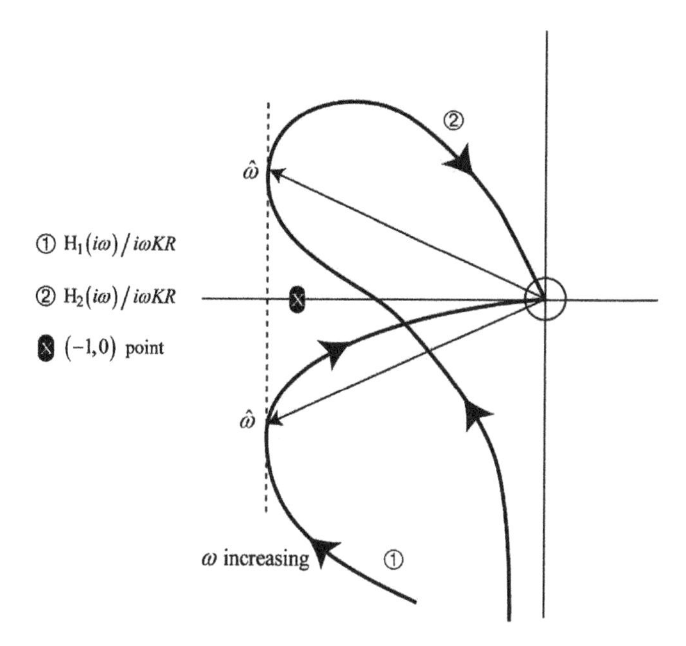

Figure 3.8 Two Stable Individual Stations; but Unstable in Parallel

Though each station is stable in isolation under coupled control, Figure 3.8 shows they could be unstable in a parallel combination. However, the depicted situation patently cannot arise if their open loop frequency responses do not cross the negative real axis. That is, the stations are each unconditionally stable.[15] Accordingly, the sufficient Grid stability criterion [80,117] devised by Butterfield et al. [150], is

> *"The Nyquist Diagrams for each station in conceptual isolation must imply unconditional and adequate stability at all output powers."*

Given a stable multi-station Grid network with totalized nameplate ratings R_T, there is the academic question—what if a conditionally stable station with coupled control and a nameplate rating R were then to be synchronized? After connection the modified open loop response is derived from Equation 2.10 and Figure 3.7 as

$$\frac{1}{R_T + R}[R_T(H_T(i\omega)/i\omega KR_T) + R(H(i\omega)/i\omega KR)] \tag{3.13}$$

[15] If an increase in the scalar controller gain causes instability, the system is termed Conditionally Stable.

where:

$$H_T(i\omega) = \sum_{k=1}^{K} H_k(i\omega);$$

$H_k(i\omega)$ – transfer function for kth coupled controlled station

and

$H(i\omega)$ – transfer function for the additional coupled station

Because for the United Kingdom

$$20\text{GW} \lesssim R_T \lesssim 60\text{GW} \quad \text{and} \quad R \lesssim 1\text{GW}$$

then by defining

$$r = R/R_T \ll 1$$

equation (3.13) approximates to

$$(1 - r)[H_T(i\omega)/i\omega KR_T] + r[H(i\omega)/i\omega KR]$$

Thus the connection of an additional coupled-controlled station affects stability margins by between $1^2/_3$ to 5%, which is negligible. If a decoupled station is synchronized, the open loop response of the Grid network becomes modified to

$$(1 - r')[H_T(i\omega)/i\omega KR_T] \quad \text{with} \quad r' \triangleq R/(R_T + R)$$

which is again negligible.

These examples so far demonstrate the value of

i. a comprehensive non-linear simulation,
ii. engineering insight and experience,
iii. the utility of Real Frequency Response functions (Nyquist diagrams) in solving complex practical problems.

Though these "working functions" evolve as more tractable SISO systems, their proper formulation is rooted in MIMO system theory.

Finally, popular UK media often comment that some particular Grid-connected wind farm can supply a certain number of homes. Such statistics often assume that *all* turbines are providing their rated maximum outputs and a daily average energy consumption of about $1^1/_2$ kW per household. Table 1.4 shows that the capacity factors of wind turbines is around 20%, so the predicted number of homes should at least be reduced by a factor of 5. During intervals in globally popular events like the World Cup, a large number of homes simultaneous brew tea or coffee and a typical electric kettle alone consumes 2 kW. In fact at the end of a 1970s Miss World competition UK power demand surged at 2 GW/min. As just described Grid operation necessitates a close instantaneous match between generated and demanded powers, and not daily averaged values. Such media statistics are therefore fallacious and suggest quite unrealistic contributions from wind energy. Adopting their same argument would suggest that 2 kW electric kettles could be properly fused on the basis of a daily averaged milliampere current.

3.4 GRID DISCONNECTION FOR A NUCLEAR STATION WITH FUNCTIONING "SCRAM"

The previous examples illustrate the practical value of linearized models in solving operational problems in both fossil and nuclear power stations. However, the large rapid power variations in accident situations invalidate linear models, and comprehensive non-linear simulations are crucial in order to provide confirmation that internal plant constraint boundaries are never breached (e.g., boiler and turbine temperature profiles). Statistics [59,65] indicate that a station's connection to its Grid network will almost certainly be disrupted during normal operation by impacts on transmission lines from large birds, aircraft or lightning. Granted a functioning reactor shutdown (scram) system, a particular accident control strategy must be devised to restrict the induced thermal stresses across the plant so that its longevity is not compromised. For example a temperature difference of some 100 °C across a steam-generator tube potentially causes rupture. This fault situation belongs to a set of so-called Design Base Accidents,[16] and the

[16] Refer to Section 4.1

following fast reactor example in Figure 3.9 demonstrates the necessity of

i. industry specific experience,
ii. a transparent one-to-one relationship between control and controlled variables, and
iii. a thoroughly validated non-linear simulation, for which the experimental data are acquired as far as possible from broadly similar plants and electrically powered heat-transfer rigs.

Because power is no longer extracted from a station's turbo-alternators after a Grid disconnection, their steam control valves (TCVs) must be abruptly closed to prevent over-speeding and consequential damage. In the assumed circumstances, a safe shutdown of the nuclear chain reaction is achieved by unlatching a more than sufficient number of gravity-driven absorber-rods. Though the chain reaction is terminated, some 6% of pre-trip reactor power is initially produced from the radioactive decay of fission products and the significant thermal energy residing in plant coolants and metalwork [117,118]. The correspondingly diminished steam production can evidently be dissipated in the station's condensers for which purpose heat transfer processes are first maintained by emergency power supplies and then by the promoted natural circulation. Thermodynamic efficiency is generally promoted by preheating boiler feed water with steam bled from a number of points along a turbine, so closure of the TCVs causes a rapid fall in feed water temperature. The sodium coolant in fast reactors has an especially large thermal capacity and provides a highly efficient mode of heat transfer [64] which in these circumstances could aggravate temperature differences across tubes of the counter-flow steam generators and Intermediate Heat Exchangers (IHXs). With a full recirculation boiler design [117,142] (Lamont), preheated feedstock from a deaerator is usually injected downwards [117] along the length of a steam drum or occasionally into downcomers, so as to make up some 10% of the boiler input. Consequently, water-inlet temperatures with this design are well buffered thereby reducing the induced hoop stresses in boiler tubing. However, economic considerations favor fast reactor stations with once-through boilers (Benson) [117] without a steam drum, so water-inlet temperatures can therefore change far more rapidly to aggravate the potential for damaging thermal shocks. Accordingly, some fast reactor

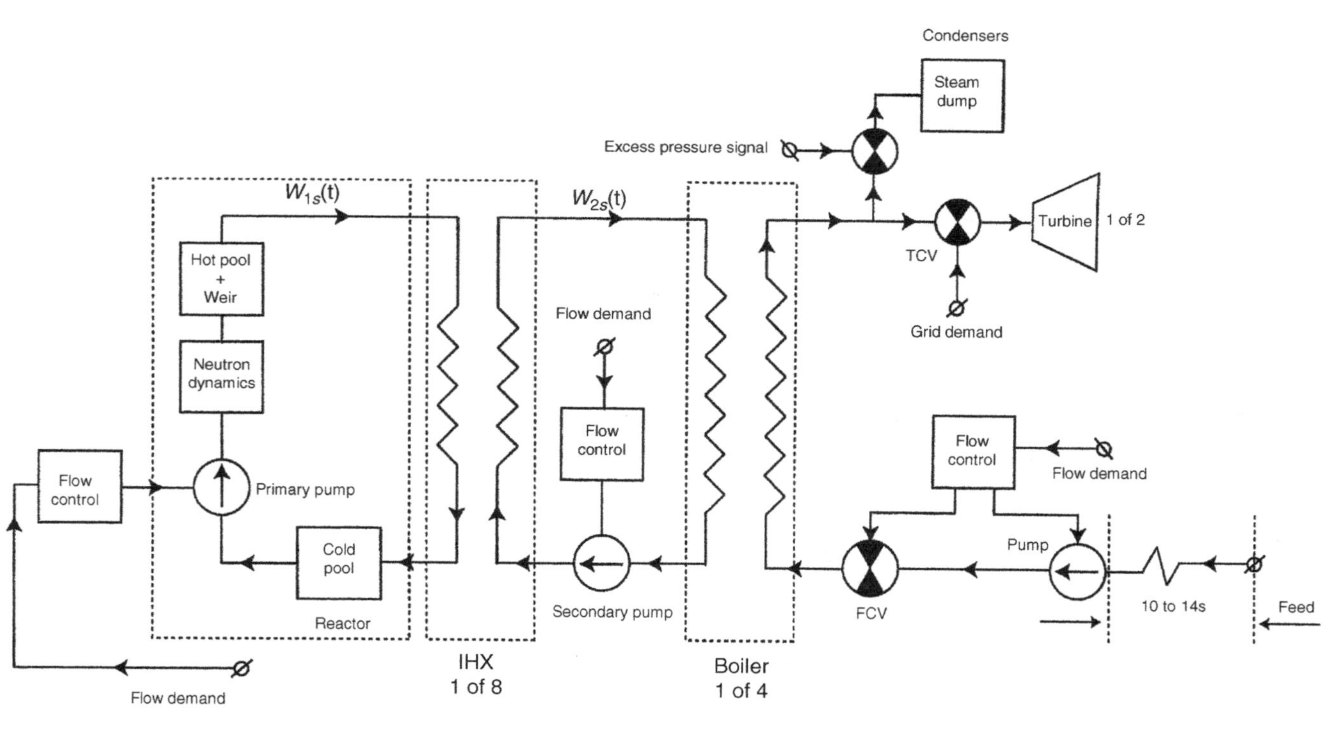

Figure 3.9 Once-Through Boiler Fast Reactor Simulation for a Grid Disconnection in a DBA at Full Load

plant designers in the United States advocate [152] an extra standby tank of preheated feed water. However, a simply implemented trip sequence for once-through boilers in fast reactor stations is now described that achieves acceptable thermal stresses without any additional capital equipment [141].

A schematic diagram for the fast-reactor system under consideration is shown in Figure 3.9, and it includes components considered for a proposed. British commercial system (CFR) with helically tubed once-through boilers.[17] The plant actually consists of one reactor, eight counter-flow IHXs, four counter-flow boiler units and two turbo-alternators to generate an electrical output of 1320 MW. For detailed non-linear digital simulation studies with the JCBARK[18] program, symmetrical operation at full-power is first investigated. A typical or average channel is modelled for the reactor, IHXs and boilers with the actual power transfers derived by straightforward linear scaling. After the TCVs and reactor scram-rods are abruptly tripped [58], it is proposed that the primary and secondary sodium pumps are operated with mass flow rate control (rather than the more usual speed) to effect

$$\left.\begin{aligned} W_{1s}(t) &= 0.1 + [W_{1s}(0) - 0.1](1 + t/12)^{-1} - \text{primary kg/s} \\ W_{2s}(t) &= 0.1 + [W_{2s}(0) - 0.1](1 + t/12)^{-1} - \text{secondary kg/s} \end{aligned}\right\} \tag{3.14}$$

Matched primary and secondary sodium flows as above prevent excessive temperature differences and thereby damaging hoop stresses in IHX tubing. Over-pressurization of the steam generators is mitigated by steam dumping [142] which as a percentage w_D of the full-load value is according to

$$\left.\begin{aligned} &w_D = 0 \quad \text{for} \quad P \leq 165\,\text{bar} \\ &w_D = 80(P - 165)/8.75 \quad \text{for} \quad 165 \leq P \leq 173.75\,\text{bar} \\ &w_D = 80 \quad \text{for} \quad P > 173.75\,\text{bar} \end{aligned}\right\} \tag{3.15}$$

[17] CFR was never built, as attention moved to a European design (EFR) which also never materialized.

[18] A simulation of Jointed Construction for Boiler and Rig Kinetics, which was subsequently extended to include reactor, steam drum, feed train, and steam turbine modules. It was developed by Drs. A. Robins, D. Farrier and the author several years earlier.

Due to preheated water held initially in the feed trains and deaerators, the inlet water flow can be maintained for 10 s before colder liquid begins to enter the boiler unit with the shortest feed main. At this point in time τ_{feed} the water feed pumps' set point and control valves are switched to match the natural recirculation rates for the primary and secondary sodium circuits. The form of the waterside heat-transfer correlations suggests that this artifice broadly attenuates temporal rates of boiler-tube temperature changes by the inverse ratio of the pre-trip water-inlet flow rate to that existing any time thereafter. In addition to mitigating thermal stresses, the proposed shutdown strategy attempts to conserve density disparities between "hot and cold legs" of the sodium circuit so as to encourage natural circulation.

Simulated inlet and outlet temperatures for the IHXs in Figure 3.10 and the inlet temperatures of a steam generator in Figure 3.11 confirm the effectiveness of the proposed accident-control procedure for the full-load situation with $\tau_{feed} = 10$ s. Corresponding normalized waterside inlet and outlet flows in Figure 3.12 clearly reveal the closure of the TCVs, the opening of the steam dump to the station's condensers, and the feed water flow switch at 10 s. Individual feed mains vary in length, and those of the PFR at Dounreay corresponded to a delay τ_{feed} of 10–14 s. Robustness of the proposed control strategy is demonstrated by the predicted temperature transients in Figures 3.13 and 3.14, with the longest feed-water flow switch at 14 s. Though temperature changes at a boiler inlet have somewhat larger excursions

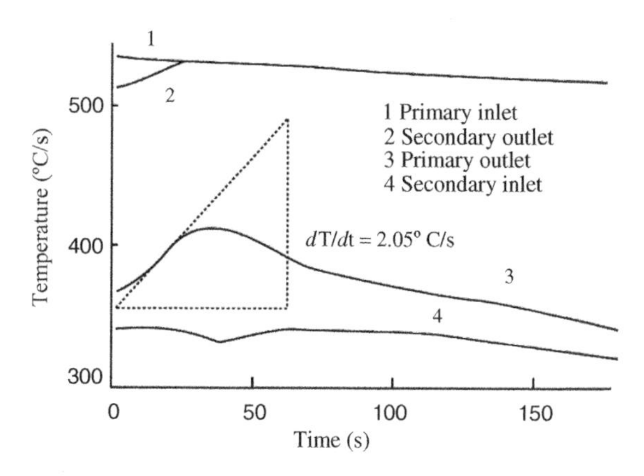

Figure 3.10 IHX Temperatures $\tau_{feed} = 10$ s

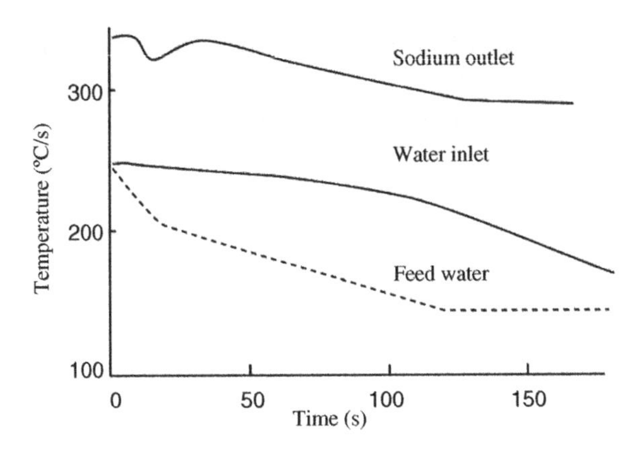

Figure 3.11 Waterside Temperatures $\tau_{\text{feed}} = 10\,\text{s}$

than before, they remain acceptable and IHX temperatures are hardly affected. Because radioactive decay power decreases with pre-trip reactor power and time, a full-load trip appears as the worst-case scenario [59]. Additional simulation results (not shown) for a range of load factors and $\tau_{\text{feed}} = 14$ s confirm this inference as well as the desirability of identical primary and secondary sodium-pump run-down rates.

Industry-specific experience and a thoroughly validated non-linear simulation have been shown to be essential for devising normal and

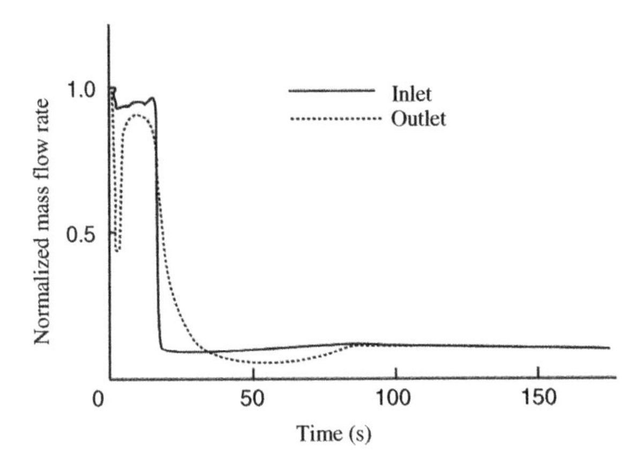

Figure 3.12 Normalized Boiler Mass Flows $\tau_{\text{feed}} = 10\,\text{s}$

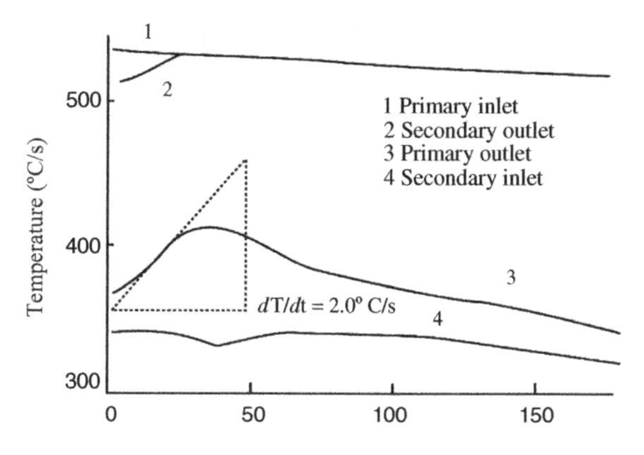

Figure 3.13 IHX Temperatures $\tau_{\text{feed}} = 14$ s

accident control strategies in nuclear power plants at the preconstruction stage. With this simulation available, Nyquist techniques using real frequency responses are the obvious choice for designing controllers for normal maneuvres. Personal experience and this example also suggest that ad hoc accident-control strategies cannot be humanly conceived if a controlled variable were to be a function of several control variables (i.e., via a scalar matrix). The interventions of reactor safety trip-systems and acceptable rates of thermal change intrinsically impose markedly different response times for the controlled variables in nuclear and fossil-fired power stations. These result in the desirable one-to-one

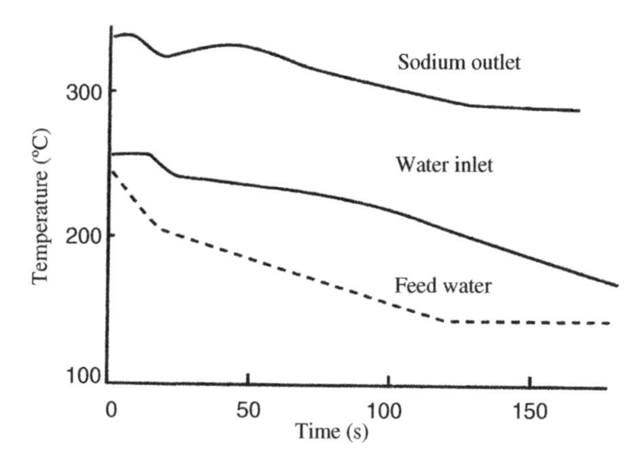

Figure 3.14 Waterside Temperatures $\tau_{\text{feed}} = 14$ s

relationships between the control and controlled variables as illustrated in Sections 3.1,[19] 3.2, and 3.3. Widely different response times were also engineered in early self-adaptive systems for oil-well drilling [153], missiles [154] and communication receivers [155] in order to create more tractable SISO control problems.

[19] Given an electrically powered feed pump.

CHAPTER 4

Some Aspects of Nuclear Accidents and Their Mitigation

4.1 REACTOR ACCIDENT CLASSIFICATION BY PROBABILITIES

The robustness of an engineering system is limited by economic factors in the sense that expenditure is only justified in making the plant adequate for its intended purpose. Failure to recognize this principle results in the equipment losing its market or not being built, through either excessive price, or on the other hand lack of reliability and performance. Where a catastrophic failure in components could be the precursor to loss of life, and a nuclear power station is not unique in this respect,[1] reliability and performance necessarily include the safety of the plant's operators and of the general public. Under these circumstances, the statistical risks to life and to the environment must be quantified, provided with ranges of uncertainty, and compared with other risks present in everyday life. Such an unambiguous scientific approach orients designers toward definite goals, identifies "weak links" in a proposed system, and most importantly establishes quantitative criteria for a decision-making process [156].

[1] Refer to Section 1.3 regarding failure of the Banqiao Dam.

Nuclear Electric Power: Safety, Operation, and Control Aspects, First Edition.
J. Brian Knowles.
© 2014 John Wiley & Sons, Inc. Published 2014 by John Wiley & Sons, Inc.

The various fault conditions of a nuclear power station may be broadly classified in terms of the probabilities of their occurrence [59,65]. Historically the distinction made is according to

$$\text{DESIGN BASE ACCIDENTS with a typical aggregate probability} \leq 3 \times 10^{-4} \text{ per operating year} \quad (4.1)$$

$$\text{SEVERE OR OUTSIDE DESIGN BASE ACCIDENTS with a typical aggregate probability of} \leq 10^{-7} \text{ per operating year} \quad (4.2)$$

Because of the considerable operating experience with conventional plant items like feedpumps, turbines and electrical distribution systems, the probabilities for many design base accidents can be specified with relatively small uncertainties. From a mathematical or philosophical view, data exists from which a relative frequency based estimate of a failure probability can be made in an a posteriori sense. On the other hand, Severe Accidents can involve events outside direct experience (e.g., fuel–coolant interactions involving tonne quantities). In this case, failure probabilities must be assessed a priori with relatively larger uncertainties by an informed judgment of the available indirect evidence. These two probabilities are patently quite different in concept, and in fact form the basis of the Venn and Bayesian philosophies [159,160]. However, by regarding the experience of plant component failures and current scientific knowledge about the underlying physical processes in a Severe Accident as information that allows judgments to be refined, it is possible to pursue a unified conceptual approach [176]. Thus all probabilities in nuclear power plant risk assessments may be considered as Bayesian in concept and having the familiar combinatorial properties.[2]

Design Base Accidents (DBA) usually originate from the failure of a single comparatively inexpensive component. On the other hand Severe Accidents which were described as hypothetical before Three Mile Island have relatively low probabilities because they generally[3] originate from multiple failures of massive static structures and/or plant protection systems whose reliability is deliberately enhanced by

[2] This viewpoint is adopted in Reference [57].

[3] An aircraft impact on the containment is one exception; see Missile Studies Section 6.4.

redundancy. With these observations in mind, typical design goals may be related to the probability of a fault's occurrence [59,65]. Thus normal operational fault transients with a probability of greater than 3×10^{-4} per year are required neither to prejudice the design life of a plant nor progress beyond the failed item. Restricted progression of damage is permissible for less probable faults within the Design Basis, but there must be only a minimal release of fission products. From these more serious design base faults designers select by past experience, certain so-called "limiting events"[4] for which the adequacy of safety systems is confirmed by experiments and digital simulations. All other events within the design basis are then argued to have less serious consequences (e.g., lower fuel temperatures, higher coolant flows, etc.) than the limiting events, thereby establishing the safety of the plant for Design Base Accidents. Section 3.4 illustrates the ad hoc methodology behind the design of DBA control strategies.

Severe Accidents in nuclear power plants are characterized by the melting of a significant part of their fuel inventory. Some 98% of a plant's radionuclides are locked away in the fuel's crystal lattice [65] and the actual amounts increase with operational life (burn-up). Melting allows their accrescence for a potentially large atmospheric release ($\gtrsim 1\, kCi$), and to protect the locality licensing regulations after Three Mile Island typically stipulate [59,65,91,108]

1. The aggregate probability of all Severe Accidents must be no greater than 10^{-7} per operating year.
2. The most exposed individual is subject to a quantified not unreasonable hazard. At Three Mile Island this person received less than 20% more than the natural background [66] dose[5] of 2–3 mSV.
3. The number of addition cancers expected is demonstrably very much less than the normal incidence [157].
4. The obligatory preparation of a well-conceived evacuation plan for plant staff and public. Fatalities at Three Mile Island were caused by road accidents in the panic exodus.

[4] A double-offset shear-break of a cold leg of a PWR with all emergency cooling systems functional is one example of such a limiting event. US federal regulations require statements of modelling assumptions and risk assessment criteria.

[5] All exposed to a radiation hazard wear an obligatory film badge or monitor while on duty.

5. The mandatory simulator training of plant staff and a designated hierarchy of responsibility based on professional skills and experience.

Measures to achieve these objectives will be outlined below. The dangers posed by some reactor fission products will be addressed next.

4.2 HAZARDS FROM AN ATMOSPHERIC RELEASE OF FISSION PRODUCTS

Catastrophic breaches of a reactor vessel and its reinforced concrete containment by an MFCI, a disintegrating plant fragment or a hydrogen explosion are circumspectly engineered to be highly improbable. From all precursors the aggregate probability is typically no greater than 10^{-7} per operating year, or assuming a Poissonian distribution there is an expectation of one such rare event in 10 million years. However, granted such an event, the release of fission products would pose the principal danger to public health and the environment. In this context, the hazard is represented by the expected increased[6] number of cancers induced in the surrounding population. The fission products from U-235 and Pu-239 are quite similar, and toward the end of a 3-year fuel cycle the actual fission product inventories of fast and thermal reactors are broadly the same. Consequently the hazards from both reactor types are similar. With regard to plutonium it is neither a significant chemical poison nor a radiological hazard because

i. Its principal emission is α-particles which pose no threat outside the human body.
ii. The half-lives of Pu-239 and Pu-240 are about 24,000 and 6500 years respectively [76], so the radiation dose per unit of absorbed mass is relatively low.
iii. Though inhalation constitutes the most serious hazard, retention in the lungs occurs in common with other aerosols for sizes 1 to 5 μm only. Particles below 1 μm tend to be exhaled while those above 5 μm are expelled in phlegm.

[6] Those additional to the natural incidence.

Releases of some nuclides like Strontium and Caesium rapidly plateout, so serious effects on public health can be prevented by isolating the surrounding area from the food chain for a limited period of time. Noble gases like Krypton and Xenon form a significant portion of the fission product inventory, but their hazard is markedly reduced because there is negligible lung absorption of these gases. On the other hand wind-borne aerosols of radioiodides or their chemical salts[7] are readily absorbed, and this effect is aggravated by a selective accumulation in the thyroid. For these reasons, radioiodides[8] or their compounds are generally considered [161,162] to be the major hazard to public health in the unlikely catastrophic failure of a reactor and its containment system. Indeed, Farmer's Reactor Safety Criterion [157] is based on the likely number of additional thyroid cancers if the iodides in a fission product inventory were to be released at ground-level. Developments in digital computer hardware and simulation techniques now allow dose-rates and induced cancer statistics to be calculated for many components of a fission product release and with more detailed representations of local population density and dispersion due to weather conditions. Even so, the predictions still over-estimate the hazards (grossly so according to some opinions [104]) by neglecting important alleviating factors. For example

1. With a water reactor, the mass of iodides available for dispersion would be markedly reduced by dissolution in the large quantities of water and vapor in the reactor and its containment [104]. With fast reactors, their sodium coolant would operate in a similar manner.
2. It seems likely that fission products would be released into the atmosphere as a hot rising plume. This effect, which promotes their dispersion and lowers the dose rate, is not represented in some computer simulations.
3. The hazard is defined in terms of additional induced cancers, whereas owing to highly effective surgery actual fatalities for the thyroid condition are only about 10 to 20% of all presentations [163,164].

[7] Current evidence suggests that caesium iodides are the forms present in a fuel element, but their stability on release depends on temperature and the reducing or oxidizing character of the environment [104].

[8] Principally the β and γ emitter I-131 with a half-life of about 8 days.

4. Data on radiation-induced cancers is available only from the high-dose cases of Japanese A-bomb survivors. Predicting the incidence of cancers caused by an atmospheric release of reactor fission products necessitates a linear extrapolation to much lower dose-levels. However, no observable human health effects have been demonstrated below about 20 mSv, due almost certainly to the regular natural replacement of body cells. Indeed, the Chernobyl exclusion zone is now populated by normal healthy families of wolves and bears [322]. This linear dependence can therefore be reasonably assumed to be pessimistic [163].

4.3 MATHEMATICAL RISK, EVENT TREES, AND HUMAN ATTITUDES

The Mathematical Risk R_A corresponding to an event A is defined as

$$R_A = P_A H_A \tag{4.3}$$

where P_A and H_A denote respectively the probability and the hazard associated with A. For example if A denotes a particular Severe Accident, then H_A would be the likely number of cancers from the quantity of fission products released, P_A the probability of A per annum and R_A the expected incidence of radiologically induced cancers per year. Though the above definition reasonably maintains a constant risk as the hazard increases in severity but its probability correspondingly decreases, unquantifiable human aspects frequently complicate a risk assessment. Certain forms of death for instance are regarded with greater repugnance than others, and in such cases there is an instinctive demand for an indefinably smaller risk. Hence, the risk from a Severe Accident must be related to the natural incidence of thyroid cancer in particular. Risk of death from other causes are of course useful as a means of placing this event in perspective, and some recent statistics are given in Table 4.1. It shows that our reactions to externally and self-imposed risks are markedly different, and this constitutes an apparently intractable facet of risk presentation to the general public. For example, the likely number of thyroid cancers caused by the Three Mile Island-2 accident are shown in Section 4.4 to be orders of magnitude less than deaths from the natural incidence, yet as shown by Table 4.1 people

Table 4.1
Certain Risks to the UK Population

Type of Event	Natural Thyroid Cancer Deaths in 2008 [164]	Fatal Road Accidents in 2009 [165]	Deaths Due to Smoking in 2009 [166]
Annual Total for the UK Population	354	2222	>81,400

campaign vociferously against the dangers of nuclear power while quietly accepting relatively much larger self-imposed risks. In addition, by involving just the tractable issues of plant operability and public health, risk assessments for nuclear power have become unbalanced. Dunster [168], for example, asserts that "If we are wishing to make a judgement about the merits for being an energy-consuming society, we must consider not only the risks of generation but also the benefits." During a televised interview [169], Jonas Salk[9] similarly commented, "We are so often preoccupied with the dangers to our society that we tend to overlook the opportunities."

Event tree methods are being increasingly used by manufacturers and licensing authorities in the presentation of all types of safety cases. Fundamentally, this graphical technique involves the construction of flow diagrams like those used in computer programming. Each branch as in Figure 4.1 represents a mutually exclusive event that it is assigned a conditional probability to reflect both the likelihood of the event and the completeness[10] of current knowledge. A hazard value or function is also attached to each branch, which in the context of a Severe Accident represents the additional mass of radioiodides released by the event. The joint probability, hazard and risk of any final or intermediate fault condition are then calculated from a tree by visual inspection. Apart from simplifying calculations, event trees provide a systematic unambiguous presentation of probable accident sequences and serve to highlight those posing the dominant risk. In this way, they identify improvements necessary in the safety features of a design and suggest the most cost-effective implementation.

[9] Discoverer of the poliomyelitis vaccine.

[10] E.g., 0.1 ± 0.001 indicates a very well-understood phenomenon.

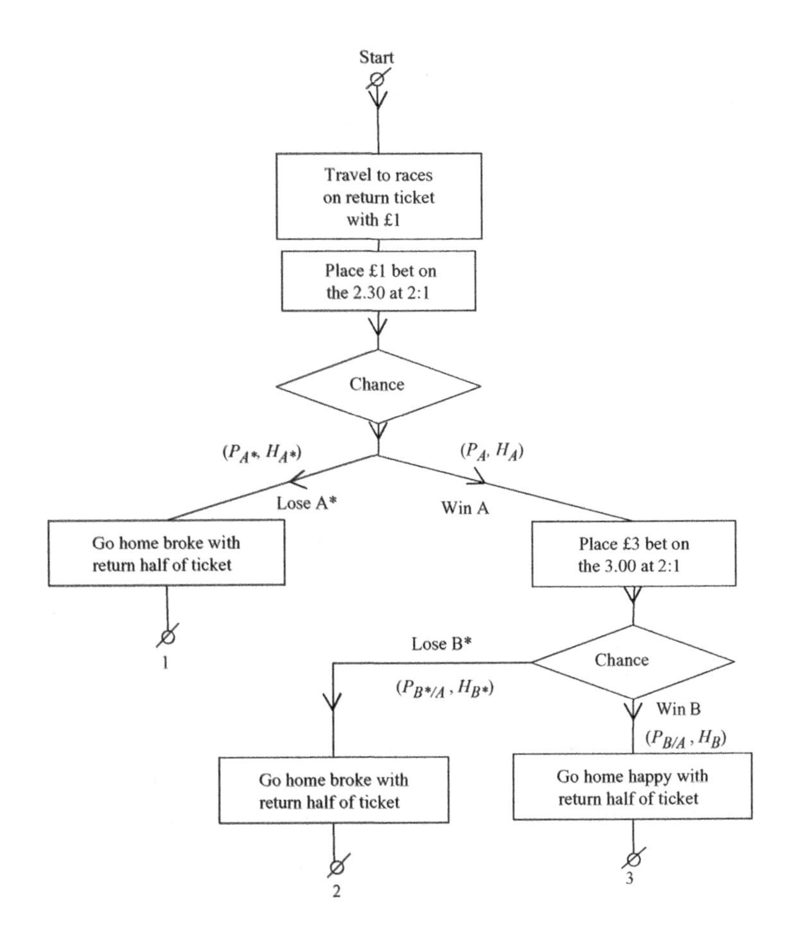

P – Probability, H – Hazard, R – Mathematical risk

A*, (B*) – Not A, (B)

$$P_3 = P_{B/A} \cdot P_A = P(A,B), \quad H_3 = H_A + H_{B/A}, \quad R_3 = P_3 \cdot H_3$$

Figure 4.1 A Careful Punter's Event Tree

Using upper bounds for the conditional probabilities and risks of accident events, Farmer [157] published the first probabilistic risk analysis for the siting of Advanced Gas Cooled Reactors. The later and much more comprehensive Reactor Safety Studies by Rasmussen [167]and the Federal German Risk Study [65,97] are also notable for specifying the spreads on probability estimates. However, the US

Reactor Safety Study [167] omits the important role of operators in either alleviating or exacerbating an accident situation. Furthermore, it assumes that accidents lead exclusively to either an assured cooling of the core or a "melt-down" with an inevitable breach of the containment. The later Three Mile Island-2 accident clearly demonstrates the importance to safety and risk assessments of both of operator responses and of a partially degraded core yet intact containment situation. Accordingly, extensive theoretical and experimental research on degraded core situations was subsequently assembled on a multinational collaborative basis. As justified to some extent by the following discussions, these investigations principally centered around

i. Robustness of fuel cladding to extreme reactivity insertions or coolant flow reductions [77].
ii. Coolability of a degraded core both inside and outside the reactor vessel [93,94,100,181,182].
iii. Rupture of a reactor vessel by the shock mechanical loads [88,102,103] or missiles created by internal explosions (fuel–coolant interactions [86,89,90,146]).
iv. Rupture of the containment by missiles [68,105,106] from external sources, or by hydrogen explosions in the particular case of a water-cooled reactor.
v. Formation and propagation of aerosols [104,170,171].
vi. Passive safety systems exploiting natural circulation [108,109].

In Section 4.4 below, Farmer's Criterion [157] quantifies these requirements as do later recommendations.

4.4 THE FARMER-BEATTIE SITING CRITERION

The probability per annum $P(C)$ that a reactor accident releases C curies[11] of radioiodides is derived from the usual general form [173] as

$$\frac{dP(C)}{dC} = g(C) \tag{4.4}$$

[11] 1 curie $= 3.7 \times 10^{10}$ Bq; 1 Sv $=$ 100 rem and 1 man-rem is an individual dose of 1 rem.

where $g(C)$ is the so-called probability density function of C. Its primitive is the probability distribution function of C with the properties

$$P(C_2) - P(C_1) = \int_{C_1}^{C_2} g(C)dC$$
$$= \text{Prob. per annum of a release between } C_1 \text{ and } C_2 \tag{4.5}$$

and

$$P(C) = \int_{-\infty}^{C} g(C')dC' = \text{Prob. per annum of any release} \leq C \tag{4.6}$$

Because the scales of $P(C)$ and C are so wide in reactor safety assessments, Farmer and Beattie [157] adopt logarithmic scales for both with

$$\frac{dP(C)}{d(\log_{10} C)} = f(C) \tag{4.7}$$

and as

$$\frac{d}{dC}(\log_{10} C) = 1/2.303C \tag{4.8}$$

then

$$g(C) = f(C)/2.303C \tag{4.9}$$

Equations (4.4) and (4.9) reveal the physical dimension of $f(C)$ as $(\text{years})^{-1}$, so its ordinates are often referred to as $(\text{years})^{-1}$ or those of $[f(C)]^{-1}$ as years. Substitution of equation (4.9) into (4.5) yields

$$P(C_2) - P(C_1) = \int_{C_1}^{C_2} [f(C)/2.303C]dC \tag{4.10}$$

If C is the median point between C_1 and C_2 lying on $f(C)$ with

$$C_1 \triangleq C/\sqrt{10} \quad \text{and} \quad C_2 \triangleq C\sqrt{10}$$

then equation (4.10) evaluates [157] as

$$P\left(C\sqrt{10}\right) - P\left(C/\sqrt{10}\right) = \alpha f(C) \quad \text{where} \quad \alpha = 1.576 \tag{4.11}$$

Reactor safety assessments provide a spectrum of Severe Accidents with varying ground-level concentrations of the principal radioiodides. A straight line with gradient $= -1$ can then be drawn in $\log_{10} P(C)$ and $\log_{10} C$ coordinates to represent an acceptable upper bound in terms of the criteria (4.2) and (4.3) on page 81, for all the investigated cases. Though points on such a line correspond to equal risk as defined by equation (4.3), they do not represent equal fatalities because lower absorptions favor our bodies' natural repair processes and curative surgery. Also because adverse public concern and reaction increases with an increasing hazard, an arbitrary 3/2 weighting is adopted for the bounding line [157].

$$F(C) \triangleq AC^{-3/2} \quad \text{for} \quad C \geq 1\,kCi \tag{4.12}$$

Over the years worldwide commercial reactor operations have accumulated several thousand operational years. Within this context, what constitutes a publically acceptable risk? Farmer [157] argues that the release of less than 1 kCi in 1000 years should be deemed reasonable in order to restrict lost power production and diagnostic investigations to sensible proportions. Because the number of Severe Accidents is obviously a discrete variable and because such accidents are engineered to be rare events, statistical characterization by a Poissonian distribution [173] is the most appropriate. Accordingly, the probability of n iodide releases of 1 kCi in T years is in general given by

$$P(n) = \frac{(\upsilon T)^n}{n!} \exp\left(-\upsilon T\right) \tag{4.13}$$

where υ is the expected number of events per year. For an expected release of just one 1 kCi in 1000 years, it follows that

$$\upsilon T = 1$$

so

$$P(1) = 0.37 \quad \text{and} \quad P(0) = 0.37$$

Farmer and Beattie [157] derive sufficiently close values of 0.33 on the less certain basis of a Normal distribution to justify the smooth transition from equation (4.10) to

$$F(C) \triangleq 10^{-2} \quad \text{for} \quad C \lesssim 50\,Ci \tag{4.14}$$

in Figure 4.2, which is called the Farmer Curve. Its sufficiency as an upper probability bound for an acceptable radioiodide release in the United Kingdom, Severe Accidents are now considered in the context of criteria (4.2) and (4.3) on page 81.

If $h(N)$ denotes the usual form of probability density function for the number of thyroid cancers presenting as a result of a radioiodide releases, then similar to the above

$$\frac{dP(N)}{dN} = h(N) \quad \text{and} \quad P(N) = \int h(N)dN \tag{4.15}$$

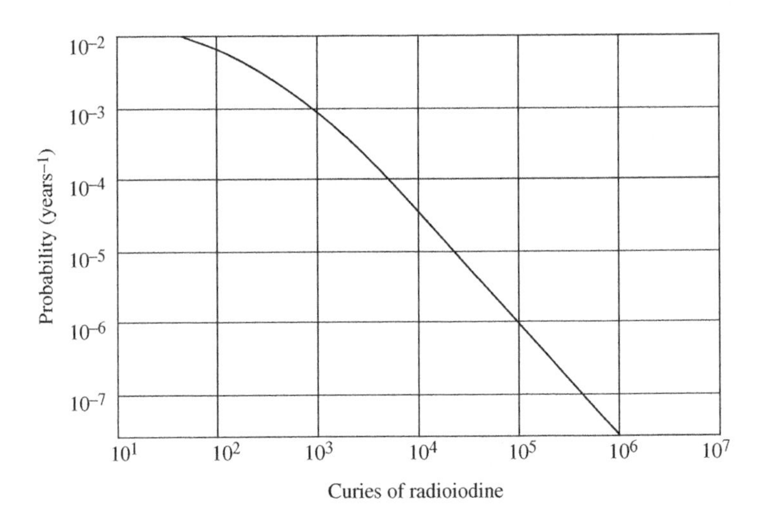

Figure 4.2 Farmer's Bounding Probability Curve [157]

and on the more convenient logarithmic scales

$$\frac{dP(N)}{d(\log_{10} N)} = H(N) \quad \text{with}\, h(N) = H(N)/2.303N \tag{4.16}$$

with:

$$P(N_2) - P(N_1) = \int_{N_1}^{N_2} [H(N)/2.303N]dN \tag{4.17}$$

Though the Lebesgue Measure [114,173] is strictly required to accommodate the probability density and distribution functions of both discrete and continuous variables, δ-functions [124] and Riemann Integration provide a more accessible and tangible appreciation of the above. Due to the complex natures of weather patterns and population density an evaluation of equation (4.17) is intractable in practice, so some simplifying yet conservative approximations are necessary. In this respect UK statistical data on wind velocities[12] and Pasquill's six weather categories for aerosol dispersion are averaged to produce a wind factor W from which the dose D rem from an emission of C Curies is described by

$$D = WC/r^{1.5} \tag{4.18}$$

where r is the radial distance from a source. The expected number N of thyroid cancer patients presenting out of Q who receive this dose is then approximated by

$$\left.\begin{aligned} N &= QBD \quad \text{for}\, D \geq 10\, rem \\ &= 0 \quad\quad\;\; \text{for}\, D < 10\, rem \end{aligned}\right\} \tag{4.19}$$

where $B = 15 \times 10^{-6}$ per rem is the mean of age-dependent values[13] specified by the International Committee on Radiological Protection

[12] Clearly dependent on location.

[13] Very conservative; see statement (4-2-4).

[174]. A representative population of 4 million around an AGR is assumed to be uniformly distributed in an annulus of radii ½ and 10 miles. The inner radius indicates that few homes are usually near a site boundary, and the outer limit an intrinsically decreasing dose with distance (r). Wind directions are quantized into 30° sectors which are also the limits of aerosol dispersion.

With the above conservative assumptions the expected increase in thyroid cancer presentations after a UK Severe Accident can be inferred. Calculations with the STRAP code [175] for a 10 kCi release give a total individual dose of 2.2×10^6 man-rem for the above population, and therefore 33 presenting cases per million (i.e., 15×2.2). However, for a reactor satisfying Farmer's Curve in Figure 4.2 the probability of this event is no greater than 0.66×10^{-4} per operating year, so the expected annual number of presenting cases is no greater than 0.0022 per million. Now the UK population in 2011 was around 62 million so the natural annual fatalities from the disease is derived from Table 4.1 as 5.7 per million, which thanks to surgery is an 80–90% reduction of the 1770 per million annual presentations. This calculation and others by Farmer and Beattie show that the additional annual risk to the local population from a Severe Accident with a UK reactor is exceedingly small by comparison with the natural cause: criterion (4.3).

Minimizing individual risk often figures significantly in public health criteria. With a 10 kCi release of radio-iodides the probable dose to a child at 1000 yards (914 m) is computed [175] as 400 rem. If this dose were actually received, the expectation of developing a thyroid cancer would be $6 \times 10^{-3}\left(\text{i.e., } 15 \times 10^{-6} \times 400\right)$. Ignoring prevailing wind effects, the assumed 30° sectors of aerosol dispersion imply only a 1 in 12 chances that this dose is actually received, so the risk reduces to 5×10^{-4}. Moreover, the probability of this event for a site satisfying Farmer's Curve is no greater than 0.66×10^{-4} per operating year, so the expectation of thyroid cancer for this child is just 3×10^{-8} per annum. The aggregate risk for a reactor with 4 or 5 mutually exclusive [173] releases on or close to the bounding curve is evaluated as approximately 1×10^{-7} per operating year. Thus subject to meeting the Farmer–Beattie constraint the additional annual expectation of developing thyroid cancer for an individual most at risk is orders of magnitude less than the natural incidence: criterion (4.2). Table 4.1 reveals that the self-imposed risk of death from a road accident or tobacco smoking is decades greater still.

The above analysis enables the Three Mile Island reactor accident to be placed in perspective. Although its fuel contained between 3 to 5 million curies of radioiodides [104], almost all were dissolved in the water or vapor within the sump and auxiliary building, and just 16 curies were released into the atmosphere [66]. Earlier destructive tests within small experimental reactors confirm the effectiveness of water or sodium coolants in reducing atmospheric releases of fission products [104]. The maximum additional individual exposure at TMI and that to the surrounding population have been estimated [66] at 80 m rem and 1.5 m rem respectively, whilst the natural background radiation is about 100 m rem. In this context the natural background radiation in the granite city of Aberdeen is some three times that of London which is built on Mesozoic geology: yet there is no statistically significant difference in attributable cancers for the two populations.[14]

4.5 EXAMPLES OF POTENTIAL SEVERE ACCIDENTS IN FAST REACTORS AND PWRs WITH THEIR CONSEQUENCES

Fast and thermal reactors with the same fuel burn-up have very similar fission product inventories, but the most likely causes of their atmospheric release in Severe Accidents are radically different. In a pool-type fast reactor, the core and all primary circuit components are contained within two strong nested tanks, which can be isolated from the secondary sodium pumps and steam generators by fast-acting valves. The primary sodium coolant at near atmospheric pressure provides an enormous heat sink for the decay heat (PFR $\sim$ 1 GJ/°C), which is also extracted in an emergency by a thermal-syphon[15] heat exchanger [314]. These engineered safety features are considered capable of eliminating the possibility of overheating the fuel, if an actual loss of coolant were to occur in the reactor circuit. Severe Accidents in fast reactors therefore principally concern the following initiating events [177]

[14] This is evidence of our body's self-repair mechanisms, and the pessimism of a linear extrapolation of A-bomb casualty statistics.

[15] A natural convective heat exchanger with a NaK primary side and an atmospheric secondary side.

i. Gross power excursions as induced for example by the multiple mis-replacement of breeder rods by fuel pins (despite warning instrumentation); and then followed by failure of the automatic shutdown system.
ii. Loss of coolant flow to all subassemblies as a result of failures in primary pumps, pipes or ducts; and then followed by failure of the automatic shutdown system.
iii. Loss of coolant flow to a single subassembly followed by failure to effect a reactor-trip through the burst-pin detection system, or its outlet temperature measurement, or its boiling noise detection system.

It is relevant to examine in more detail the mechanism by which melting in a single subassembly can lead to a major release of a fast reactor's fission product inventory. If for a particular subassembly there is an excessive burn-up of the fuel or a gross mismatch of gagging or a sufficiently high gas content in the sodium, then local overheating of the pins over a short time-scale would allow the ejection of molten fuel into the sodium. As described in Chapter 5, heat transfer between the two liquids can then potentially take the form of an explosive rate of vapor generation that redistributes the remaining fuel pins with a marked reduction of the interstitial sodium. By sufficiently reducing neutron absorptions in this way, the core could become prompt critical: thereby melting a major portion of its fuel with the subsequent possibility of an explosive vaporization of the liquid sodium coolant. This thermally driven explosion with molten fuel and liquid coolant is known as Molten Fuel–Coolant Interaction (MFCI). Its clear importance to fast reactor safety motivated worldwide research [88,146] which included that at AEEW. Similarly powerful explosions are observed with molten fuel and water, so the phenomenon was also investigated as part of the UKAEA water reactor safety program [89,185].

The superior economics of light water nuclear reactors and reasons for the wider adoption of PWRs rather than BWRs are outlined in Section 1.8. These arguments justified the construction during 1987–95 of the United Kingdom's first PWR at Sizewell. Any proposed nuclear power plant in the United States must be shown to meet the generic safety criteria of its Nuclear Regulatory Commission (NUREG 0737) [91], whereas in the United Kingdom a safety case must satisfy the

Nuclear Installation Inspectorate for each individual plant.[16] Accordingly, though the Sizewell plant is in essence a Westinghouse Standard Nuclear Unit Power Plant System (SNUPPS) like in Figures 1.3 and 1.4, various design modifications [178,179] are incorporated to meet the particular UK licensing requirements for normal operation, maintenance and the mitigation of Severe Accidents. Specifically lower radiation doses for plant operatives are achieved by reduced concentrations of Cobalt[17] in control-valve seatings and boiler tubing where Inconel 690 has replaced the original Inconel 600. Moreover improved radiological shielding of major plant items, remote or robotic maintenance and more alkaline water chemistry contribute to fulfilling the ALARP radiation exposure criterion.[18] Reactor scram is actioned by the proven AGR system of Laddics and physically independent self-validating μ-processor units [127].

Because the reactor coolant in a PWR is under high pressure (15.5 MPa), leaks or fractures in primary circuit pipe work or the pressurizer or the single-skin pressure vessel have an expertly assessed aggregate probability of around 10^{-4} per operating year [59,65]. A double offset shear-break of a pipe carrying inlet coolant would create an extreme large loss of coolant accident (LLOCA). On the other hand, small breaks of 2–80 cm^2 in the above components constitute a small loss of coolant accident (SLOCA), which by allowing the primary circuit to remain longer at higher pressures delays the intervention of the emergency core cooling systems (ECCS). Accordingly, as shown in Table 4.2, SLOCAs are expertly adjudged [65] as the more probable precursors of Severe Accidents. To reduce this risk the Sizewell reactor has four[19] inlet coolant nozzles and an enhanced ECCS. Specifically

i. Four larger pre-pressurised accumulators of aqueous boric acid, so that two rather than the standard three are sufficient for reactor shutdown.

[16] Satisfactory reactor siting, as described in Section 4.4, involves local population density, weather patterns, and seismic activity.

[17] Neutron absorptions create Co-60, which is a penetrating γ-emitter with a half-life of 5 years.

[18] As Low As Reasonably Practical.

[19] Earlier PWRs like Three Mile Island have just two.

Table 4.2
Probabilities of Various LOCAs as Precursors to Severe Accidents (Ref. 65)

LOCA Type	Leak X-Section (cm^2)	Probability $(year)^{-1}$	Probability of Causing a Severe Accident $(year)^{-1}$
Reactor coolant pipe			
- Large leak	>500	$<10^{-8}$	5.0×10^{-7}
- Medium leak	200–500	$<10^{-8}$	2.0×10^{-6}
- Small leak	80–200	3.1×10^{-7}	5.7×10^{-5}
- Small leak	2–80	3.7×10^{-6}	
Pressurizer			
- Transient opening of relief valve	20	8.2×10^{-7}	9.0×10^{-6}
- Unwarranted opening of safety valve	40	2.2×10^{-6}	–
Others			
- Connection line to annulus	–	$<10^{-7}$	3.0×10^{-8}
- Steam generator tube	1–12	1.1×10^{-6}	–

ii. Four high-pressure pumps (11 MPa) one of which, rather than the standard two, is sufficient to cope with a SLOCA.

iii. Two extra low-pressure pumps (2.7 MPa) for cooling major plant items including the depressurized reactor, whose heat is rejected into a closed secondary circuit of sea water or a forced draught cooling tower. Other unique engineered safety features include extra redundancy for the pressurizer[20] relief valves, and the addition of reinforced concrete to surround the usual steel containment structure.

In the event of a LLOCA, the reduction in primary circuit pressure automatically effects scram, and rupture of accumulator check values allows the injection of borated water into the PWR core.[21]

[20] Refer to Table 4.2.

[21] After a public hearing in 1970, the US Atomic Energy Agency's prescribed ECCS measures were considered in their report to make a LLOCA as generally the limiting accident.

Table 4.3
Unrestricted Progression of a Severe Accident in a PWR

Surface Temperature of Fuel (°C)	Observable Phenomena
700–750	Burnable poison rods soften, and the Cd–In–Ag content of Inconel–clad control rods melts
800	Fuel pins balloon and burst
900	Exothermic Zr–H_2O reaction starts, accelerating the rate of fuel-temperature rise
1300–1500	Formation of liquid Inconel–Zircalloy eutectic
1400–1500	Urania–Zircalloy reaction and control-rod cladding melts
1850–1950	Zircalloy melts
2400–2650	Zirconia and Urania–Zirconia mixtures melt

Furthermore, decreasing coolant flow in the core increases voidage and thereby a loss of neutron energy moderation that reduces power to decay heat values even without scram. Removal of decay heat continues with the injection of more borated water into the reactor's inlet nozzles by high-pressure pumps. After a further loss of pressure and the now open pressurizer relief valves, low-pressure pumps augment heat removal by recirculating water collected in the primary containment's sump. Cold aqueous sprays and hydrogen recombiners mitigate over-pressurization of the building from flashing coolant or large hydrogen burns. If despite the considerable redundancy in the ECCS compound failures were to allow inadequate cooling over a protracted period,[22] then decay heat would initiate the Severe Accident sequence in Table 4.3. If the meltdown were to continue then all water except that shielded by the lower-core and lower-support plates in Figure 1.4 is expected to be vaporized. These plates would support a growing accumulation of semi-solidified core debris whose outer solidified crust would result from radiant and ablative

[22] Calculations indicate no less than 1 hour, so there is time for emergency reconfigurations of a plant [65,93] or even attaching civilian fire pumps. For the fast reactor situation, refer to Reference 213.

heat transfer. Eventually, a loss of creep strength [96] in the lower core-plate would allow quantities of corium into the lower head giving the potential for an explosive MFCI that could rupture the reactor pressure vessel. However, experiments at AEEW show that the explosive energies released under such "fuel rich" conditions are markedly reduced probably due to

i. A shortage of coolant restricting the formation of a detonate-able coarse mixture (see Section 5.1)
ii. A reduced inertial constraint allowing less durable heat transfer.

Fuller descriptions of Severe Accident scenarios with Event Trees appear in References 59 and 65, but the quantities of molten fuel becoming available for an MFCI powerful enough to breach a reactor pressure vessel were not quantifiable in the 1980s. Accordingly, contemporaneous reports by the Sizewell B Committee [59], Sandia National Laboratory [97] and the Gesellschaft fűr Reaktorsicherheit [65], ascribe the wide Bayesian probability range of 10^{-1} to 10^{-4} per year for this destructive event. On the other hand, a Swedish government report [98] even denies its occurrence by virtue of the above mitigating factors and the efficacy of safety systems. Indeed post-incident investigation at Three Mile Island revealed a porous in-vessel debris bed of 8 to 16 tonne which had passively equilibrated rather than detonated. Thus even a late restoration of cooling appears to prevent an MFCI by increasing the viscosity of the molten debris: see Section 5.1. Moreover, world-wide operational legislation [108] now prescribes frameworks for operator command structures and training that render Severe Accidents far less likely. However, uncertainties in the Sizewell B report [59] persisted about sufficiently powerful MFCIs as a result of the 1965 analysis by Hicks and Menzies [85]. Assuming an isentropic (lossless) expansion of an explosive MFCI vapor bubble they predict bounding efficiencies of up to 30% for the conversion of heat into mechanical work. If such values were actually to be true, then the safety cases for fast and light water reactors would be compromised when not unrealistic quantities of molten core materials are involved. However, experiments [88,89,185] at AEEW with kilogram quantities of molten urania consistently gave conversion efficiencies with sodium or water coolants of around 4 to 5%, but scaling this value to tonne-sized reactor quantities was unacceptable due to the absence of an underlying physical mechanism. During 1990–92, the SEURBNUK-EURDYN-BUBEX simulation in Chapter 5 predicted conservative efficiencies[23] of 4 to 5% by representing condensation at a liquid–vapor interface. Identification of this thermodynamic irreversibility allows a sound extrapolation of experimental values to fast and thermal reactor scales so materially consolidating their safety cases.

A large rupture in a PWR's pressure vessel after fuel melting would allow fission product aerosols, fuel debris, steam and hydrogen from oxidation of Zircalloy clad to enter the reinforced concrete containment. Its potential fracture by over-pressurization or a hydrogen explosion could patently allow an environmental release of radioactivity. Accordingly, hydrogen recombiners and doped cold-water sprays are activated at an over-pressure of about 2 bar, and these also dissolve fission products [104]. Accelerating plant fragments (missiles) could also be created from an MFCI so the concrete structure must be engineered strong enough to withstand their impact and in addition those from crashing aircraft [59,65]. Chapter 6 describes some experimental and analytical research at AEEW which addresses these issues. Though the human and environmental tragedies of Fukushima are harrowing, a positive outcome is that nuclear power stations can be engineered to successfully withstand a major earthquake of Richter scale 9.

A large quantity of core debris (corium) falling onto the containment's floor raises concern about a core-concrete interaction [181,182]. Despite the large quantities of hydrogen gas created, concomitant steam concentrations and hydrogen combiners inhibit detonations [65]. Calculations [65,182] also suggest that the building's structural integrity would be preserved by a limited atmospheric venting at 6 bar over-pressure and water injection from the sump or external means.[24] While German studies [65] provide no evidence that extended melting through the floor could be avoided (the so-called China Syndrome), such melt progression did not actually occur at Chernobyl. Following a breach in a PWR's pressure vessel, a high concentration of

[23] By virtue of the complex physical processes and uncertainties about physical properties, Severe Accident calculations are required to be only reasonably conservative, rather than the ±10% for normal engineering design.

[24] For example, civilian fire pumps [65].

aerosol particles would initially exist in the containment building's steamy atmosphere. However, they would be rapidly deposited by attaching themselves to the remaining 95% of non-radioactive ones or condense on fixed surfaces: thereby reducing the spread of environmental contamination [59,65,104,171,183].

CHAPTER 5

Molten Fuel Coolant Interactions: Analyses and Experiments

5.1 A HISTORY AND A MIXING ANALYSIS

A potential explosion when molten reactor fuel mixes with its vaporizable coolant is an example of a more general phenomena. Outside the nuclear industry other highly destructive thermal detonations involve iron + water [186,187]; aluminum + water [188]; liquefied natural gas + seawater [189], soda ash + water [190]; coal tar + water [191]; and glass + water [192]. However, the sparseness of reports indicates that such explosions are far from frequent events. Indeed, if they were other than rare, the corresponding industrial processes would now have been discontinued. Moreover, many technical recordings show molten lava from on-land or subsea volcanoes mixing passively with seawater, and the last documented detonation was at Krakatau in August 1883. This rarity in usually passive industrial and natural processes clearly indicates that special conditions are necessary for a detonation.

Though true scientific investigation began in the nineteenth century [186], the physical processes underlying an explosive interaction only became partially understood after 1960 [193]. Even now, however, knowledge of some aspects remains incomplete. From pioneering

Nuclear Electric Power: Safety, Operation, and Control Aspects, First Edition.
J. Brian Knowles.
© 2014 John Wiley & Sons, Inc. Published 2014 by John Wiley & Sons, Inc.

experiments with aluminum and water in 1970 Laker and Lennon [194] suggest that the formation of a "macro-mixture" is a necessary precursor for an explosive interaction. Then a localized disturbance, such as the partial collapse of a coolant vapor film around the molten metal, triggers a propagating detonation.[1] They (correctly) conjecture that the shock wave would escalate in intensity by finely fragmenting the melt to achieve the rapid heat transfer rate[2] for the detonation timescale of around 1 ms. Finally, though hydrogen production is recorded, later experiments by Robinson and Fry [192] prove the oxidation of aluminum to have been irrelevant. The necessity of first creating a particular coarse-mixture morphology of the two liquids is confirmed by subsequent experiments with kilogram quantities of molten metals or ceramics. Available evidence establishes that a potentially explosive coarse mixture has a sponge-like structure with a cell size of around 10 mm. Figure 5.1 depicts part of a typical untriggered[3] coarse mixture

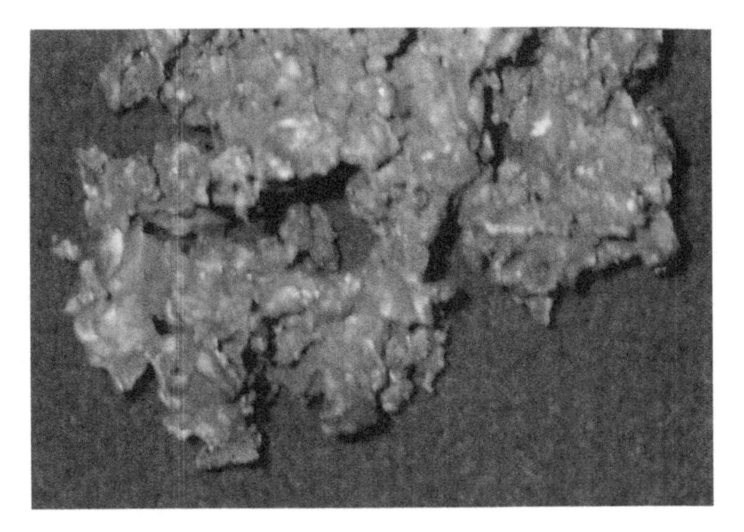

Figure 5.1 A Coarse Mixture Formed in Aluminum–Water Experiments

[1] A detonation (rather than a deflagration) is characterized by a pressure wave spatially in front of a reaction [203]. "Pinking" in petrol engines is a well-known example of an unwanted detonation rather than a deflagration.

[2] By increasing surface contact areas and decreasing heat diffusion distances.

[3] Refer to Section 5.6.

recovered from an aluminum–water experiment at AEEW. The following simplified fluid dynamics analysis confirms the necessity for first creating an appropriate coarse mixture over a timescale significantly longer than the 1 ms of a detonation.

If a one-step division process over τ seconds transforms a contiguous mass M into identical spheres each separated by their own diameter d, then the work done against the surrounding liquid drag forces is of order [195]

$$E_{F1} \simeq \frac{3}{4\rho d}\left(\frac{M}{\tau}\right)^2 \tag{5.1}$$

with

ρ—coolant density.

The thermal energy change of the initially molten corium is

$$E_C \simeq MC_p\Delta T \quad \text{with} \quad C_p \simeq 500\,\text{J/kg} \tag{5.2}$$

where ΔT is the temperature difference in the corium prior to and then after a postulated interaction. Computer simulations of Severe Accidents suggest that a fast reactor provides the extreme molten corium temperature of 5000 K, so it is assumed here that

$$2500 \lesssim \Delta T \lesssim 4000\ ^\circ\text{C} \tag{5.3}$$

with:

$$\rho_{\text{sodium}} \simeq 800\,\text{kg/m}^3 \quad \text{and} \quad \rho_{\text{water}} \simeq 1000\,\text{kg/m}^3 \tag{5.4}$$

As described in Section 5.7 the diameter of a fragmented corium particle must be below about 250 μm for its energy to be transferred over the 1 ms timescale of a detonation. Over this restricted size-range experiments provide the approximate mass mean diameter

$$d \simeq 100\,\mu\text{m} \tag{5.5}$$

Granted the total elapsed time of an MFCI event as 1 ms, the above equations give

$$\left.\begin{array}{ll} 3.75\,\mathrm{M} \lesssim E_{F1}/E_C \lesssim 6.0\,\mathrm{M} & \text{water coolant} \\ 4.70\,\mathrm{M} \lesssim E_{F1}/E_C \lesssim 7.50\,\mathrm{M} & \text{sodium coolant} \end{array}\right\} \tag{5.6}$$

Hypothetically, if the entire 100 tonne fuel inventory of a PWR were to be involved, it follows that

$$3.8 \times 10^5 \lesssim E_{F1}/E_C \lesssim 6.0 \times 10^5 \text{ in a PWR} \tag{5.7}$$

and similarly for a 20 tonne fast reactor fuel load

$$0.94 \times 10^5 \lesssim E_{F1}/E_C \lesssim 1.5 \times 10^5 \text{ in a fast reactor} \tag{5.8}$$

On the other hand, consider a two-stage process in which a contiguous mass M is first mixed into spheres of diameter D in 1 s, and then finely fragmented into spheres of diameter d in τ second. The approximate number N of larger spheres is evidently

$$N \simeq \left(\frac{6}{\pi D^3 \rho_f}\right) M \quad \text{for a fuel density } \rho_f \simeq 10^4\,\mathrm{kg/m^3} \tag{5.9}$$

and their mixing energy requirement is obtained from equation (5.1) as

$$E_M \simeq \left(\frac{3}{4\rho D}\right) M^2 \quad \text{with} \quad D \simeq 10\,\mathrm{mm} \tag{5.10}$$

Similarly the energy E for finely fragmenting just one sphere of diameter D in τ second is

$$E \simeq \left(\frac{3}{4\rho d}\right)\left(\frac{\pi}{6} d^3 \rho_f\right)^2 \frac{1}{\tau^2} \tag{5.11}$$

Equations (5.9)–(5.11) yield the two-stage fragmentation energy as

$$E_{F2} \simeq \frac{3M}{4\rho}\left[\frac{M}{D} + \left(\frac{\pi}{6}d^3\rho_f\right)\left(d\tau^2\right)^{-1}\right] \tag{5.12}$$

and from equation (5.2)

$$E_{F2}/E_C \simeq \frac{3}{4\rho\, C_p\Delta T}\left[\frac{M}{D} + \left(\frac{\pi}{6}d^3\rho_f\right)\left(d\tau^2\right)^{-1}\right] \tag{5.13}$$

For the range of temperature differences in equation (5.3) and a total PWR fuel inventory of 100 tonne, equation (5.13) evaluates for a 1 ms detonation as

$$3.8 \times 10^{-3} \lesssim E_{F2}/E_C \lesssim 6.0 \times 10^{-3} \text{ in a PWR} \tag{5.14}$$

and for a 20 tonne fast reactor fuel inventory

$$0.93 \times 10^{-2} \lesssim E_{F2}/E_C \lesssim 1.5 \times 10^{-3} \text{ in a fast reactor} \tag{5.15}$$

Though the above are merely "scoping calculations," the orders of magnitude involved indicate that an explosive MFCI with a one-step fine fragmentation process for a sizeable portion of a reactor fuel load is impossible. However, an experimentally consistent two-stage process could involve just a plausibly small fraction of the fuel's thermal energy, so MFCIs cause concern in water and fast reactor safety assessment [59]. Significantly, the coarse-mixing energy term in equation (5.13) is totally dominant which possibly explains the natural rarity of detonatable morphologies.

5.2 COARSE MIXTURES AND CONTACT MODES IN SEVERE NUCLEAR ACCIDENTS

A reliable upper bound for the fraction of detonatable mixture[4] is patently an important parameter in reactor safety assessments. In this respect the explosive power of MFCIs restricts experiments to no more than 100 kg of corium stimulants, whereas Severe Accidents could involve tonne-quantities of reactor materials. Such a wide extrapolation

[4] Hereafter simply "coarse mixture" denotes the detonatable portion of melt.

is obviously a moot point. The total heat content of a melt-mass is broadly conserved as

$$\text{Surface area of melt/thermal capacity} \approx (\text{mass})^{-1/3} \tag{5.16}$$

which implies that with increasing mass its average temperature remains higher for longer. It does not imply that the fraction of coarse mixture asymptotically approaches 1 as the total melt mass increases. Figure 5.1 and volcanic lava flows clearly reveal that a coarse mixture could not form where the melt's surface is close to or below its freezing point. In fact the formation of a coarse mixture involves many physical processes such as heat diffusion in melt and coolant, heat transfer from it by principally radiation,[5] melt viscosity and contact mode of melt and coolant. The two-dimensional CHYMES code [197] was developed to consolidate the opinion that partial freezing would impose an upper bound on the fraction of coarse mixture if large quantities of corium were to be involved. Though the computed vapor flow rates match experiments [184], thereby suggesting a reasonable simulation of an evolving potentially pre-detonatable mass, the required upper bound did not become available. Nevertheless, present knowledge enables the convincing safety assessment in Section 5.7.

The analysis in Section 5.1 and later in Section 5.7 reveal that debris sizes must be less than about 250 μm for an effective contribution to the explosive energy of an MFCI, and that these fine particles largely stem from the fine fragmentation of a coarse mixture across a shock front. On this basis the CORECT2 and THINA experiments with a sodium coolant provide a 40% upper bound on the detonatable fraction of a coarse mixture [82,206]. However, urania–sodium experiments at AEEW show that hydrogen generated by the prerequisite cleansing of the solidified debris with methanol, and then its vacuum distillation, augments the fine debris fraction.[6] Moreover, unlike a urania–water interaction, a sodium coolant experiment very often produces a number of incoherent weaker detonations [86], which plausibly augment the recovered fine debris. Accordingly, urania–water experimental data

[5] Thermal radiation flux is proportional to the 4th power of absolute temperature.

[6] Established by repetitions of this recovery procedure. Also individual debris particles are microscopically smooth with a water coolant, but they are pitted in the case of a sodium coolant, which suggests a chemical attack.

from Rig A and the Molten Fuel Test Facility at AEEW give the arguably more reliable upper bound of 20%. Figure 5.2 schematically illustrates the MFTF in which up to 20 kg of urania from a depleted uranium–molybdenum thermite mix is injected into water so as to replicate in some ways corium slumping into the residual water in a PWR pressure vessel. While the experimental and reactor contact modes are not too dissimilar, the pertinent volumetric ratios of corium to water differ by orders of magnitude. Specifically, experimental ratios

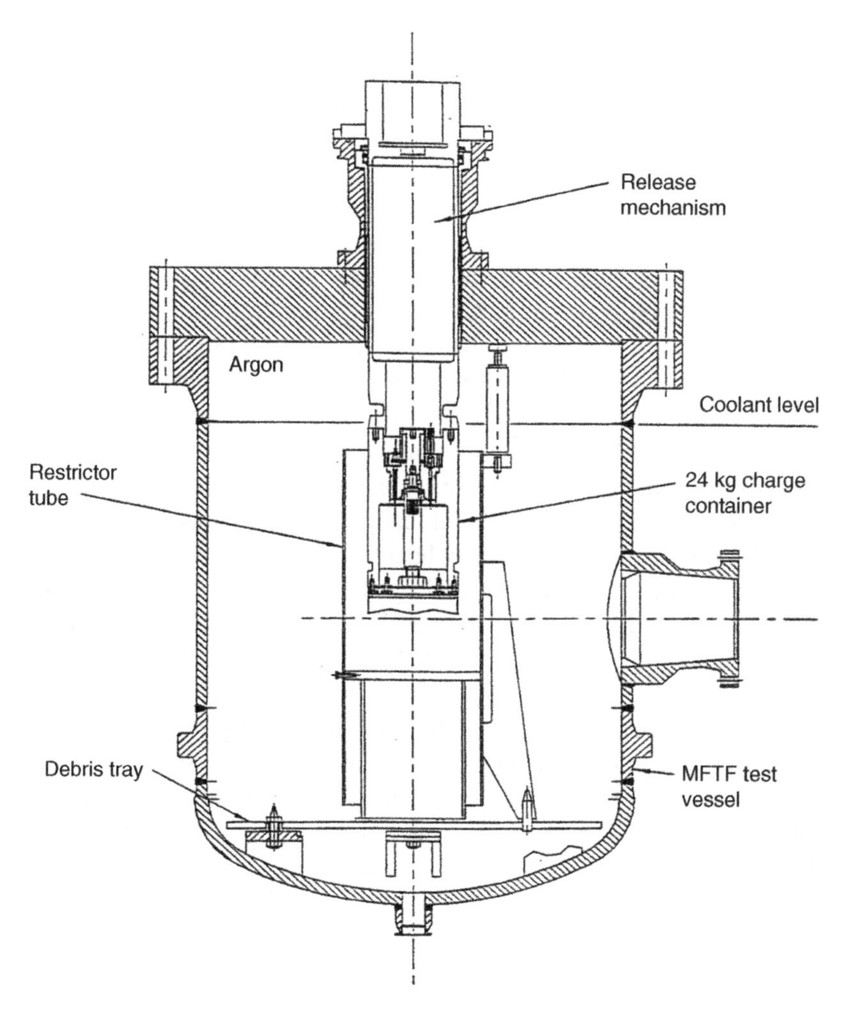

Figure 5.2 The Molten Fuel Test Facility for the SUS Experiments

are at least 1:1000 but in a reactor situation[7] the ratio can be as low as 1:30. Experimental results show that explosive energies are markedly reduced when "fuel-rich" mixtures are involved because

i. A shortage of coolant restricts the formation of a coarse mixture.
ii. A reduced inertial constraint (a tamp) allows less durable heat transfer between melt particles and coolant.

Expert opinion [59,65] identifies Severe Accidents in PWRs that could probably result in molten corium slumping into residual water in the lower head at pressures in the range 1 to 155 bar. Data on MFCI at higher than atmospheric pressure is sparse, but available evidence [98,198] indicates that with increasing pressure more violent triggers are required and that the energy release is greater. An actual reactor experiment at EG and G-Idaho observed an MFCI at the highest recorded ambient pressure of 64 bar.

Section 4.5 describes the inception of a Severe Accident in a fast reactor subassembly, in which potential MFCI might occur in a radically different geometry from that of a PWR or the MFTF in Figure 5.2. SCARABAEE [200] and TRAN [201] tests establish that the sodium content of an affected subassembly would first vaporize before any melting of the steel-clad fuel pins. At this juncture MFCI are patently impossible. Due to decay heat the steel cladding subsequently melts at $\simeq$1200 °C to be followed by the mixed oxide at $\simeq$3000 °C. Because the boiling point of steel is around 3000 °C, its vapor condenses at the cooler subassembly inlet and outlet to form strong blockages. Molten corium is then considered to perforate the subassembly wrapper and thereby allow pressurized injections of molten corium into the inter-wrapper gap or a neighboring subassembly. So far significant quantities of molten fuel would not be involved, and therefore there would be little chance of damaging escalations. However, due to fluid inertias, corium ejection would end by the development of a negative differential pressure around 2 bar which would encourage re-entry of liquid sodium. As part of a European research program [86] also involving the CORECT2 and THINA experiments,

[7] See Section 5.5.

the MFTF was modified as in Figure 5.3 to investigate this different contact mode.
These independent tests each elicited an erratic series of relatively small MFCIs. Furthermore, with the original unrestricted contact mode in Figure 5.2 just one of these SUS tests had any semblance of coherence with a principal interaction followed by a series of very much smaller ones.

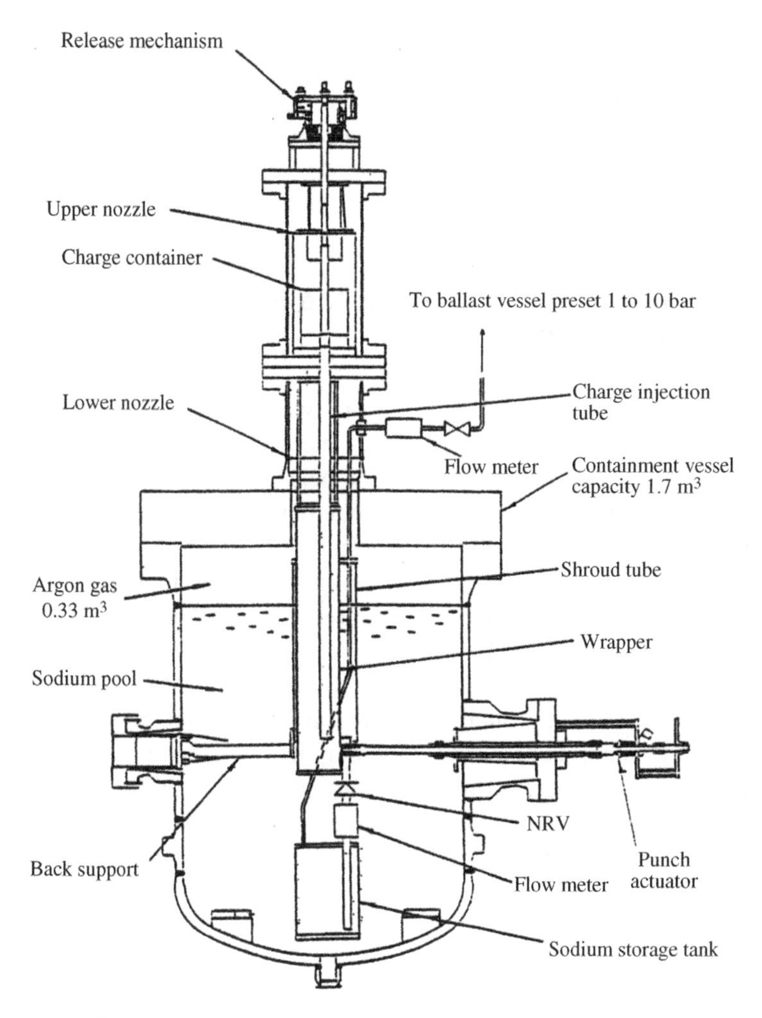

Figure 5.3 General Arrangements for B-Series Experiments

Interactions with sodium therefore appear markedly different from and far less damaging than with a water coolant. This can reasonably be attributed to the two orders greater thermal diffusivity of sodium compared to water.[8] As described in Section 5.4, this enables a much higher heat transfer rate from the melt and by promoting localized freezing obstructs an essentially coherent propagation. The most probable outcome for this particular Severe Accident appears to be a less rapid and damaging pressurization of the reactor vessel by an erratic series of small MFCIs or by slower conventional[9] heat transfer from larger corium particles (a so-called Q^*-event [202]). Finally, fast-reactor safety is enhanced by the 2000 tonne or so of sodium in its primary circuit. If 50 tonne of molten corium passively equilibrated with this coolant, the predicted temperature rise is only about 42 °C, which emphasizes that the hazard of an MFCI resides in a localized almost coherent heat transfer to a vaporizable coolant.

5.3 SOME PHYSICS OF A VAPOR FILM AND ITS INTERFACE

Leidenfrost [204] in 1756 reported that water droplets can endure for several seconds on a sufficiently hot metal surface as a result of an intervening vapor film that inhibits heat transfer. With decreasing surface temperature the vapor film could no longer be sustained and the droplet quickly boiled away. The minimum surface temperature for the existence of a vapor film became known as the Leidenfrost Point. However, later experiments and theoretical analyses in the twentieth century showed that the Leidenfrost Point is not just a property of the liquid and surface temperature. Specifically, particulate impurities or surface roughness are found to reduce the formation energy [205] of an embryonic vapor bubble and so precipitate nucleation. Indeed with extreme liquid purity, stillness and surface purity, boiling on a wetted surface can be inhibited until temperatures are well above normal and towards the limiting homogeneous nucleation temperature [205].

[8] Thermal diffusivities of liquid sodium and water are 50×10^{-6} and 0.14×10^{-6} m^2/s respectively.

[9] That is, without a shock wave.

The destabilization of a vapor film (triggering) to initiate an MFCI in a Severe Accident differs radically from the collapse of a Leidenfrost vapor film. First, liquid corium temperatures are orders of magnitude larger ($\gtrsim$3500 K) so an enormous radiant heat flux component[10] enters the liquid interface. Secondly, a film would contain non-condensable hydrogen[11] or fission product gases that can modify its thermodynamic states and thermal conductivity [3,210]. Finally, an explosive liquid to liquid heat transfer over μ-seconds is induced by external shock waves: rather than spontaneous passive nucleation over seconds. Preliminary aluminum–water experiments at AEEW confirm the necessity of a shock wave to destabilize otherwise quiescent film boiling. These tests in Rig A repeatedly created a detonable coarse mixture like that in Figure 5.1 without an MFCI. Later, a small chemical explosive was used to initiate a weak shock wave which then consistently triggered an MFCI. On the other hand, subsequent urania–water experiments in Rig A and MFTF were consistently triggered just by the injection of melt. Accordingly, to achieve greater understanding for reactor safety assessments, experimental [207] and theoretical [206] research was undertaken into shock wave destabilization of vapor films.

As can be readily visualized from Figure 5.1, a realistic three-dimensional simulation of triggering is presently intractable. Nevertheless by characterization of the pertinent physical processes useful insight can be gained from the one-dimensional model in Figure 5.4. Due to their widely different acoustic impedances a trigger pressure wave in the liquid propagates only weakly into the vapor. Also many reflections of this transmitted wave occur in the relatively thin vapor before the next stress wave in the liquid arrives back at the interface.[12] Consequently, as inferred from Figure 5.5, vapor pressure can be considered spatially uniform at any instant. Moreover, because thermodynamic relaxation times [211] are of order 1 ns and film destabilization requires [207] some 20 μs, classical thermodynamic state variables can describe each point in a film. However, by increasing fugacity [3], permanent gases in a vapor film during a Severe Accident alter these states from those of the pure substance. For example, if the partial pressure of a permanent gas is

[10] See Section 5.4.

[11] In a water reactor.

[12] See Ref. [206].

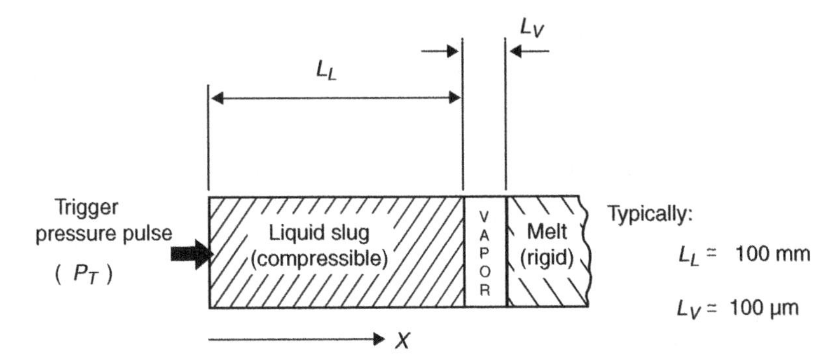

Figure 5.4 Components of the Planar One-Dimensional Model

P_0, then the saturation vapor pressure P_{sat} becomes changed by [210]

$$\delta P_{sat} = (v_{LSAT}/v_{GSAT}) \cdot P_0$$

where

v_{LSAT}, v_{GSAT} – specific volume of saturated liquid and vapor, respectively.

Table 5.1 presents the specific volume ratio v_{LSAT}/v_{GSAT} and the fraction of undissociated steam as a function of pressure at 3500 K [208,209]. Because the fission product pressure for locally rupturing a fast reactor fuel pin is of order 2 MPa, the change in the saturation vapor pressure of sodium is seen to be largely negligible. Accordingly, it is inferred that all other thermodynamic states in a film mixture remain essentially those of its separate constituents. In the case of a water reactor, the corresponding specific volume ratio is seen to increase slowly enough for the permanent gases to have a negligible effect on the saturated steam pressure: especially as increasing pressure inhibits its dissociation. It is again concluded that all thermodynamic states in a film mixture are those of its separate constituents.

The extreme temperatures in a vapor film significantly affect its local thermal conductivity. Available data [209] for superheated sodium vapor appear restricted to temperatures no greater than 1500 K. Nevertheless, it can be assumed to behave as a perfect gas at higher temperatures. On this basis kinetic theory [210] suggests a two-fold increase in thermal conductivity across a sodium vapor film to the obvious benefit of its stability. In the case of superheated steam, thermal conductivities in Table 5.2 exceed kinetic theory predictions by virtue of more mobile hydrogen molecules created by its partial dissociation.

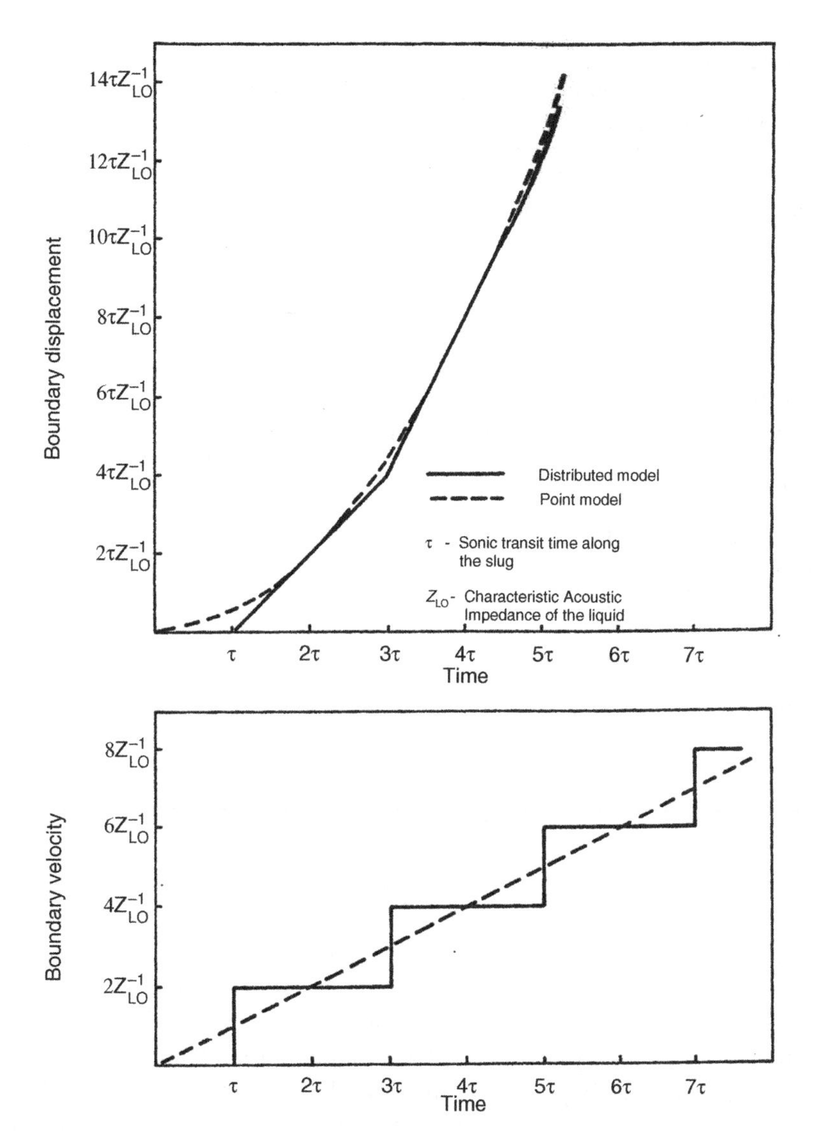

Figure 5.5 Comparison of the Distributed (Compressible) and Point (Incompressible) Models for the Kinetics of the Liquid Slug

Accordingly, film stability considerations should involve a distributed model of heat diffusion in the vapor, but too many mesh points would clearly aggravate numerical problems as a film collapses.

If water is in thermal equilibrium with its vapor at 60 °C, the evaporation and condensation mass fluxes derived from kinetic theory

Table 5.1
Specific Volume Ratios for Sodium and Water as a Function of Pressure at 3500 K

Saturation vapor pressure (MPa)	0.01	0.1	1.0	10.0
Specific volume ratio for sodium	3.7E-5	3.4E-4	3.3E-3	–
Specific volume ratio for water	7.7E-6	6.2E-4	5.8E-3	8.1E-2
Undissociated fraction of steam at 3500 K	0.03	0.24	0.57	0.79

Table 5.2
Thermal Conductivity of Steam[a] (mW/m-K)—Refs [208,209]

κ(mW/m – K) / Temperature (K)	Pressure (MPa)			
	0.01	0.1	1.0	10.0
500	–	35	37	Liquid
1000	–	93	94	102
1500	207	203	201	200
2000	423	333	296	280
2500	1960	900	553	424
3000	7450	2940	1285	751

[a]*Note*: Dissociation is suppressed with increasing pressure.

are about $2\text{kg/m}^2 - \text{s}$ or 6.7×10^{26} $\text{molecules/m}^2 - \text{s}$. Despite this chaotic interchange at an interface, experiments [214,215] show that it remains plane to within 1 or 2 molecular diameters due to intermolecular attraction. When a shock wave accelerates the liquid into its vapor, Rayleigh–Taylor waves [216,217] are suppressed so the interface remains locally plane. On the other hand if an interface decelerates close to a hot melt, these waves could conceivably distort an originally plane interface element. No analysis is apparently published that engrosses both surface tension and viscosity, but larger predicted oscillatory amplitudes would patently obtain without viscous damping. Assuming a uniform deceleration over half the typical 20 μs destabilization period [207] of a 100 μm vapor film by a 10 MPa trigger, then with just surface tension the fastest growing wavelength λ^* and its time constant τ^* are shown in Table 5.3 from calculations

Table 5.3
Fastest Growing Taylor Wavelengths and Growth Time Constants for the 20 μs Collapse of a 100 μm Steam Film by a 10 MPa Trigger

Saturated pressure (MPa)	0.01	0.1	1.0	10.0
Surface tension (mN/m)	68.5	58.8	42.3	12.1
λ^* (μm)	79.4	74.8	66.1	41.7
τ^* (μs)	3.8	3.7	3.5	3.0

[206] based on [216]

$$\left.\begin{aligned}\lambda^* &= \left[\frac{3\sigma_L}{\ddot{z}(\rho_L - \rho_G)}\right]^{1/2} \qquad \text{for} \quad 0 < \ddot{z} \\ \tau^* &= \left[\ddot{z}\left(\frac{2\pi}{\lambda^*}\right)\left(\frac{\rho_L - \rho_G}{\rho_L + \rho_G}\right) - \left(\frac{2\pi}{\lambda^*}\right)^3\left(\frac{\sigma_L}{\rho_L + \rho_G}\right)\right]^{1/2}\end{aligned}\right\} \tag{5.17}$$

where

$\ddot{z}$ – uniform deceleration of liquid phase
σ_L – surface tension of the liquid
ρ_L, ρ_G – density of liquid and vapor, respectively

Though it appears that Rayleigh–Taylor waves could grow significantly during film destabilization, the large viscous shear forces associated with such rapid micron-sized wavelengths would (in the author's opinion) strangle their growth. Moreover, experiments [207] confirm a later analysis that the collapse of a vapor film is hardly resisted, so it is therefore reasonable to conclude that a liquid-vapor interface remains essentially locally plane during its destabilization.

5.4 HEAT TRANSFER FROM CONTIGUOUS MELT

Severe Accidents would result with corium temperatures in the range [59] 3500 to 5000 K, for which Planck's Black-body radiation spectra

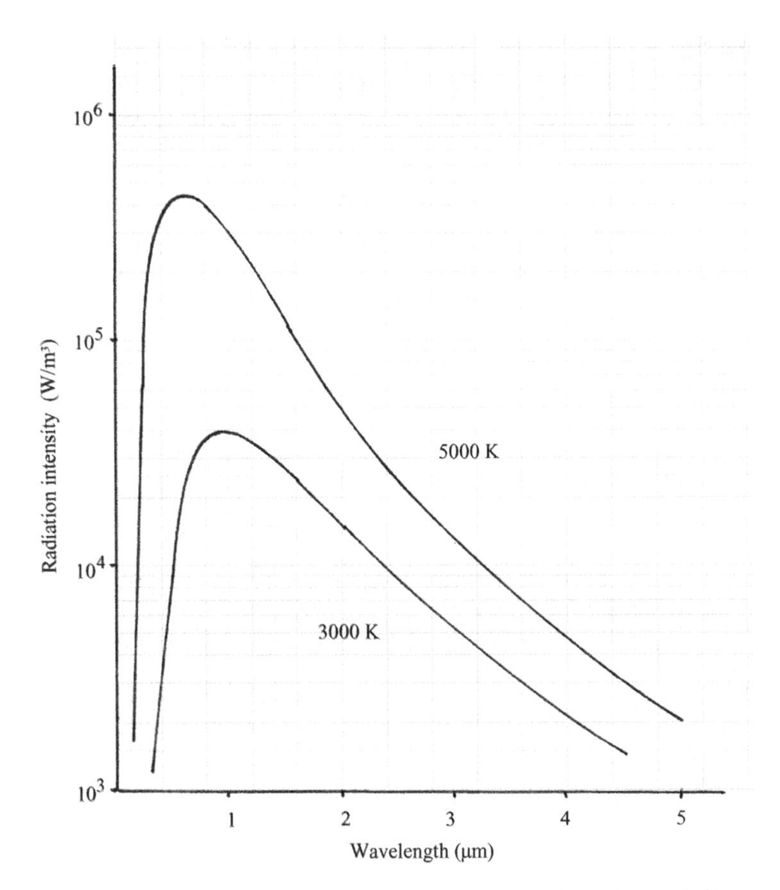

Figure 5.6 Planck's Black-Body Radiation Intensity Function (W/m^3)

[218,219] $W(\lambda)$ in Figure 5.6 has an effective waveband of $0 < \lambda \lesssim 5\ \mu m$. Similar to neutrons, the probability of an interaction between an EM wave and the atoms or molecules in a medium is found[13] to increase with its density and thickness. Accordingly, it is reasonable to postulate that the absorption of thermal radiation in a steam film depends on the product of its pressure P and thickness L. Though Doppler broadening of absorption bands also occurs with increasing temperature, the decrease in density at constant PL can be expected to predominate. With these principles Hottel [219] successfully correlates seven different

[13] X-ray studies [16] might have prompted the original conjecture.

independent data sets for the total emissivity (= absorptivity) of steam as a function of temperature by a family of curves indexed by constant values of *PL*. For example, the total absorptivity of a 100 μm thick steam film at 1 MPa is derived from his graph at $T(\mathrm{K})$as

$$a = -3.2 \times 10^{-5}(T - 1944)) + 0.055 \tag{5.18}$$

More generally, Hottel's correlation and data in Ref. [233] establish a largely insignificant absorption of thermal radiation in the steam film around an evolving coarse mixture. The total emissivity of a water surface is 0.96 [218], so it acts as a perfect absorber for present purposes. Over the waveband 4.58–6.47 μm the total emissivity of solid or liquid urania measures as 0.82 [220], and so it can be regarded as a gray emitter. On this basis the thermal radiation flux entering the interfacial liquid in Figure 5.4 is[14]

$$\phi_{rad} = \varepsilon_M \sigma \left(T_{MB}^4 - T_{LB}^4\right) \tag{5.19}$$

where

ε_M– Emissivity of $UO_2 = 0.82$

T_{MB}, T_{LB} – Interfacial temperatures of melt and liquid (water), respectively

σ – Stefan–Boltzmann constant $= 5.67 \times 10^{-8}\ \mathrm{W/m^2 - K}$

Because

$$T_{LB} \sim T_{MB}/10$$

then to a close approximation

$$\phi_{\mathrm{rad}} = \varepsilon_M \sigma T_{MB}^4 = \varepsilon_M \int_0^\infty W(\lambda) d\lambda \tag{5.20}$$

[14] Equation 4-5 in Ref. [219], which assumes the local speed of light is that in vacuo [218].

where

$W(\lambda)$ – Planck's radiation intensity function (in Figure 5.6)

The absorbed thermal radiation flux ϕ_{ab} in a slab of incremental thickness δz is generally specified by

$$\phi_{ab} = \int_0^\infty W(\lambda)A(\lambda, \delta z)d\lambda \tag{5.21}$$

where

$A(\lambda, \delta z)$ – spectral absorption factor $= 1 - \exp(-a_\lambda \delta z)$
a_λ – spectral absorption coefficient

Measured values [221,222] of $a(\lambda)$ for water at 1 bar and 20 °C are graphed in Figure 5.7. As a result of Doppler broadening, increases in temperature of over 20–60 °C decrease the resonant amplitude at 2.95 μm by around 25% but pressure has little effect [223] because water is relatively incompressible. Table 5.4 provides values of the absorption factor $A(\lambda, \delta z)$for a range of wavelengths and water thickness (μm). In the waveband 0 to 2.5 μm, less than 5% of a radiation flux

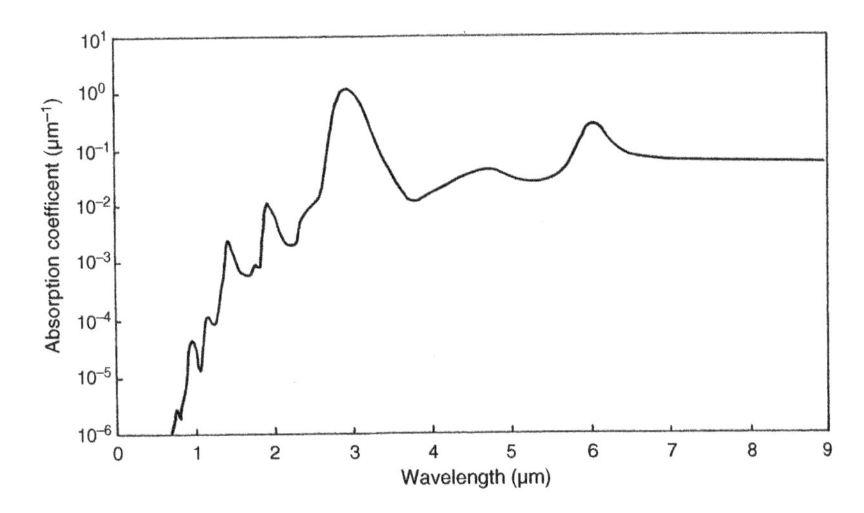

Figure 5.7 The Absorption Coefficient of Water at 20 °C and 0.1 MPa

Table 5.4
Absorption Factor $A(\lambda, \delta z)$ as a Function of Wavelength and δz in the Range 1–500 μm at Around 20 °C

λ(μm)	$A(\lambda, 1)$	$A(\lambda, 5)$	$A(\lambda, 10)$	$A(\lambda, 50)$	$A(\lambda, 500)$
1.0	3.2E-5	1.6E-4	3.2E-4	1.6E-3	1.5E-2
1.5	2.5E-3	1.2E-2	2.5E-2	1.2E-1	7.1E-1
2.0	9.0E-3	4.4E-2	8.2E-2	3.6E-1	1.0E0
2.5	1.0E-2	4.9E-2	9.5E-2	1.0E0	1.0E0

is absorbed over the first 5 μm from an interface. For corium at 3500 K over 87% of the radiated power lies in this waveband [218] with an even larger fraction at higher temperatures [59]. Though the total radiant flux at 3500 K evaluates from equation (5.20) as $7\ \text{MW/m}^2$, it follows that less than $0.3\ \text{MW/m}^2$ can be absorbed in this crucial interfacial region. By comparison, Table 5.2 reveals an average molecular conductivity for steam that is well in excess of 200 mW/m – K, so the corresponding heat flux across a typical 100 μm film with a temperature difference of at least 3000 K is

$$\phi_{\text{conduction}} \gtrsim 200 \times 10^{-3} \times \left(3000/10^{-4}\right) = 6\ \text{MW/m}^2$$

Furthermore, over an incremental time δt heat absorbed at a plane surface approximately diffuses the distance [224]

$$d \simeq \sqrt{4\,\alpha\,\delta t} \tag{5.22}$$

For water and a typical steam film destabilization period of 20 μs

$$\alpha_{\text{water}} = 0.16 \times 10^{-6}\ \text{m}^2/\text{s} \quad \text{so} \quad d = 3.6\ \mu\text{m} \tag{5.23}$$

Consequently the radiant heat absorbed by water beyond 5 μm can hardly influence film stability. It is therefore concluded that molecular conduction across a steam film is the effective stabilizing heat-transfer process. Several published simulations [198,225,226] of steam film destabilization omit the effect of a wavelength-dependent absorption coefficient on the radiant energy absorbed by the surrounding water.

Monatomic and symmetric molecules, like those of sodium vapor, undergo neither vibrational nor rotational transitions. Also symmetric molecules have no electrical dipole moment, so they can neither significantly absorb nor emit radiation by vibration or rotational bands. It follows that gases of such molecules are transparent to thermal radiation at low to moderate temperatures [218]. At high temperatures, however, these gas molecules can radiate or absorb appreciably by electron bound to bound or bound-free transitions. However, the asymmetric molecules of steam have all these degrees of freedom yet still absorb negligible amounts of thermal radiation. A fortiori, the same must be true for sodium vapor. For a good electrical conductor, like liquid sodium, the effective total absorption length l_{ab} can be estimated from the skin-depth equation [227] of *EM* wave theory

$$l_{ab} \simeq \sqrt{\frac{\lambda\, r}{\pi\, \mu\, c}} \tag{5.24}$$

where

λ – wavelength; r – electrical resistivity of the conductor

μ – magnetic permeability; c – speed of light in the conductor

In the case of liquid sodium, equation (5.24) reduces to

$$l_{ab} \simeq 13\sqrt{\lambda}\ \mu\text{m}$$

which shows that a radiated heat flux entering the liquid is effectively absorbed at the actual interface. Similar to equation (5.20) the thermal radiation flux entering the interfacial liquid sodium [219] is derived as

$$\phi_{\text{rad}} = \sigma T^4_{MB}\left({}^1\!/_{\varepsilon_M} + {}^1\!/_{\varepsilon_L}\right)^{-1} \tag{5.25}$$

Due to surface contamination the total emissivity of a sodium surface in Severe Accident conditions is certainly larger than the value measured [228] under clinically clean laboratory conditions as

$$\varepsilon_L = 0.05$$

The total heat flux entering and absorbed by the interfacial liquid sodium with respect to the coordinate system in Figure 5.4 is therefore

$$\phi_{LB} = \phi_{\mathrm{rad}} + \kappa_{GB}\frac{\partial T}{\partial x}\bigg|_{GB} \tag{5.26}$$

where

κ_{GB} – thermal conductivity of interfacial sodium vapor

$\frac{\partial T}{\partial x}\big|_{GB}$ – interfacial temperature gradient in the sodium vapor

5.5 MASS TRANSFER AT A LIQUID–VAPOR INTERFACE AND THE CONDENSATION COEFFICIENT

Under conditions of thermodynamic equilibrium the Maxwell–Boltzmann probability density function characterizes the velocities of ideal gas molecules. Assuming isotropic scattering [58], the mass flux of ideal gas molecules traversing one way through a conceptual plane surface is derived [210] on this basis from classical kinetic theory. Ignoring the significant inter-molecular attractions in the liquid and vapor states (i.e., the Joule–Kelvin effect), applications of this classical analysis to the liquid–vapor interface gives the net mass flux in the liquid as [241]

$$G_B = (2\pi R)^{-1/2} F(\sigma)\left[\frac{P_{GB}}{\sqrt{T_{GB}}} - \frac{P_{\mathrm{SAT}}(T_{LB})}{\sqrt{T_{LB}}}\right] \tag{5.27}$$

where

G_B – interfacial mass flux (kg/m^2 – s); R – specific gas constant

P_{GB}, T_{GB} – interfacial vapor pressure and temperature, respectively

T_{LB} – interfacial liquid temperature

σ – condensation coefficient ($0 < 0 \leq 1$)

It is recommended [229] that the molecular structure function $F(\sigma)$ takes the form

$$F(\sigma) = \frac{8\sigma}{2 - 0.798\,\sigma}\left[\frac{\gamma - 1}{\gamma + 1}\right] \tag{5.28}$$

where:

γ – specific heat ratio of the vapor $\left(C_p/C_v\right)$

Equation (5.27) correctly implies a net mass flux of zero when saturated liquid and vapor co-exist at an interface under conditions of thermodynamic equilibrium. However, under non-equilibrium conditions it loses some accuracy. Specifically, apart from the omission of inter-molecular forces, vapor molecules near an interface originate as

i. Those having just emerged from the liquid
ii. Those having diffused from a higher temperature near the melt
iii. Those reflected from the liquid surface

This motley ensemble is unlikely to be characterized by a Maxwell–Boltzmann distribution as required for the validity of equation (5.27). Moreover, the rigorous definition [3] of temperature is in the context of thermal equilibrium, so with net interfacial mass transport neither T_{GB} nor T_{LB} strictly exists. Nevertheless there appears no alternative to equation (5.27), and for simulation purposes they are taken as extrapolations of heat diffusion calculations in the two media. Under net evaporation or condensation equation (5.27) predicts an interfacial temperature jump (discontinuity) that is confirmed by experiments with liquid metals [230,231]. An early simulation of vapor film destabilization by Corradini [198] assumes both interfacial fluids to be saturated, but his later corrected model [226] reveals the marked effect of this discontinuity on computed transients.

A condensation coefficient is in essence the probability that a molecule impinging the interface enters the other phase. Mills and Saban [232] comprehensively review many published analytical derivations, but conclude that reliance is best placed on experimental data. They consider the most reliable measurements for water to be those by Nabavian [233], Berman [234] or themselves which together give

$$0.35 < \sigma \leq 1.0 \tag{5.29}$$

After reviewing 11 independent publications on the condensation coefficient for various liquid metals over the pressure range 0.001 to 1 bar, Fedorovich and Rosenhow [231] conclude that

$$0.1 < \sigma \leq 1.0 \tag{5.30}$$

Results are tightly clustered for pressures no greater than 0.1 bar, but thereafter their dispersion increases. The wide uncertainties in equations (5.29) and (5.30) are exceptional by twentieth century standards, and might well reflect the purity of the coolant. In fact experiments with a liquid metal indicate that the condensation coefficient decreases with increasing contamination [242]. Later laboratory measurements show that the coefficient assumes its maximum value of unity when the water or liquid metal is exceptionally pure [255], but in Severe Accidents a reactor coolant would be heavily contaminated. Consequently, because decreased interfacial condensation saps less energy from an expanding MFCI bubble, the least of the above values for a condensation coefficient should be used in safety simulations.

Permanent gas molecules in sufficient numbers can seriously reduce the mass flow rate from industrial steam condensers by restricting access to heat transfer surfaces [219]. As a preventative measure, deaerators are installed in the feed-trains of power station boilers, where they also provide emergency supplies (see Section 3.4). In Severe Accidents hydrogen or fission product gases might similarly be expected to reduce interfacial condensation rates, and thereby conserve the energy of an expanding MFCI vapor bubble. However, fast reactor tests [228] show that collapse times of sodium bubbles are largely unaffected when contaminated with Xenon concentrations representative of spent fuel. Other laboratory experiments [236,237] with steam bubbles establish that condensation rates are reduced by less than 10% with the introduction of 15% molar concentrations of nitrogen. In both cases, the authors conclude that efficient turbulent mixing must exist within a collapsing bubble. No experiments concerning the effect of permanent gases on triggered vapor-film destabilization were found in the literature. From Table 5.2 the principal effect of hydrogen is seen to increase a film's thermal conductivity and therefore to a degree its stability.[15]

The above discussion describes the considerable uncertainties in predicted interfacial condensation rates. Nevertheless, Section 5.8 shows that conservative interfacial condensation rates largely account for the experimentally observed reduction in MFCI energies from the idealized isentropic Hicks–Menzies yields [85]. From the viewpoint of reactor safety assessments, the identification of a physical process to justify the extension of experimental 4 to 5% MFCI efficiencies to reactor-scale tonne-quantities is highly significant.

[15] See Section 5.6.

5.6 KINETICS, HEAT DIFFUSION, A TRIGGERING SIMULATION, AND REACTOR SAFETY

Liquids possess elasticity as well as mass, so the interfacial liquid in Figure 5.4 does not move in unison with the application of a trigger pulse at its far end. Such lack of concomitance is often irrelevant, but here a typical experimental film destabilization period [207] of 20 μs is comparable with the 67 μs transit time of a pressure pulse. Consequently as illustrated in Figure 5.5, a distributed model of shock propagation is necessary, though some simulations [21,225,226] feature only point models. Moving boundaries often pose additional degrees of analytical difficulty, which are aggravated here by heat and mass transport phenomena. However, according to equation (5.23), heat diffuses only 3.5 μm during film collapse whilst the pressure shock travels[16] about 0.3 m. Moreover, the mass of a 100 μm of steam film is totally negligible in relation to that of the 100 mm liquid slug. Thus heat diffusion and shock dynamics in the liquid can be advantageously decoupled (i.e., solved independently). Like other moving boundary problems, one-dimensional shock propagation is best formulated in terms of the "constant mass packet" Lagrangian equations

$$\frac{\partial V}{\partial t} = -\upsilon^{o}\frac{\partial P}{\partial x} \quad \text{(momentum)} \tag{5.31}$$

$$\nu = \nu^{o}\frac{\partial z}{\partial x} \quad \text{(mass)} \tag{5.32}$$

$$\frac{\partial I}{\partial t} = -P\frac{\partial \nu}{\partial t} \quad \text{(energy)} \tag{5.33}$$

$$V = \frac{\partial z}{\partial t} \quad \text{(kinetics)} \tag{5.34}$$

[16] The isentropic sonic speed of 1500 m/s is slightly slower than the weak anisentropic shock propagation of a trigger.

$$P = P(I, \nu) \quad \text{(thermodynamic state)} \tag{5.35}$$

where

z – particle displacement along the x-coordinate axis (m)
V – particle velocity (m/s); P – pressure (Pa)
ν – specific volume (m^3/kg); I – specific internal energy (J/kg)
ν^o – unshocked specific volume

An experimental correlation relates the wave speed $\dot{x}_s$ to particle velocity and for water this takes the form

$$\dot{x}_s = \beta_0 + \beta_1 V + \beta_2 V^2 \tag{5.36}$$

Published data [208,209] shows that the sonic speed c with

$$c^2 \triangleq \left.\frac{\partial P}{\partial \rho}\right|_{S\ \text{constant}} \quad : \quad S - \text{entropy} \tag{5.37}$$

is strongly dependent on temperature but only weakly so on pressure, and therefore similarly for β_0, β_1, and β_2. Because thermodynamic equilibrium is rapidly regained across a shock front, the Rankine–Hugoniot equations [203] closely approximate pressure and energy changes. Accordingly, close to a Rankine-Hugoniot curve the relationship between pressure and internal energy is found to be represented [237] by

$$P(\upsilon) = P^o + \mathcal{G}(\nu)(I - I^o)/\upsilon \tag{5.38}$$

with the appropriate constant value for the Gruneisen function $\mathcal{G}$.

Integration over the moving mesh points in the liquid and vapor is effected by Leibnitz's theorem [112] to provide first of all ordinary differential equations, and then the required difference equations [206]. A necessary condition for the von Neumann Stability of the adopted explicit solution scheme is [208]

$$N_{CFL} \triangleq \text{Shock front speed} \times (\delta t/\delta x) < 1 \tag{5.39}$$

where N_{CFL} is the Courant–Friedrichs-Levy Number. Because triggers evolve as relatively weak shocks their propagation is approximately sonic, so for water around STP

$$N_{CFL} = 1500(\delta t/\delta x) < 1 \tag{5.40}$$

The water slug approaches the (assumed) rigid melt according to the calculation

$$z_k^{n+1} = z_k^n + V_k^{n+1}\delta t\,; \quad z_k^n \triangleq z(k; n\delta t) \tag{5.41}$$

and numerical breakdown is prevented by progressively reducing the time step with decreasing film thickness. Though explicit solutions [203,238] of the Lagrangian equations without added viscosity and thermal conductivity[17] lead to the progressive sharpening of a shock front, and then often to computational failure, the adopted solution scheme appears stable.

Neglecting relaxation effects [210,211], Fourier's heat conduction equation and the first law of thermodynamics describe one-dimensional heat diffusion in an isotropic semi-infinite slab by [224]

$$\frac{\partial T}{\partial t} = \alpha\frac{\partial^2 T}{\partial x^2} \quad \text{with } \alpha - \text{thermal diffusivity} = \kappa/\rho C_p \tag{5.42}$$

Corresponding central-space and backward-time linear difference equations for a fixed or moving mesh have tridiagonal structures which are solvable by Gaussian elimination, but preserving second-order spatial accuracy at the boundaries requires special care [117,206]. Necessary and sufficient conditions for this solution procedure to be stable with a fixed Eulerian mesh are [238]

$$\alpha\delta t/(\delta x)^2 < {}^1\!/_2 \tag{5.43}$$

Low liquid compressibility and the relatively small mass of vapor ensure that matrix terms for the moving mesh are largely those for a fixed mesh.

[17] The Lax–Wendroff solution [238] scheme introduces artificial viscosity and thermal conductivity to achieve numerical stability.

Consequently, equation (5.43) is adjudged apposite for present purposes. During the time step of a shock-wave calculation, the thermal penetration distance in water evaluates from equation (5.22) as only 0.8 μm. Using this guide, sensibly converged thermal diffusion calculations are obtained with the lattice parameters

$$\delta t = 1\,\mu\text{s}\,;\, \delta z = 1\,\mu\text{m} \quad \text{for which} \quad \alpha\delta t/(\delta x)^2 = 0.16$$

Attempted shock calculations with these values would have $N_{CFL} = 1500$, so justifying the suggested decoupling of shock and heat diffusion calculations.

If the outer surface of a semi-infinite slab of an isotropic conductor is abruptly changed by T^*, then according to equation (5.42), a temperature wave diffuses into its interior as

$$T(x,t) = T^* erfc\left(\sqrt{x/4\alpha t}\right) \tag{5.44}$$

In order that $T/T^* \simeq 0.01$, published tables [239] give

$$x/4\alpha\, t = (1.83)^2$$

Molten urania has a thermal diffusivity of order[18] $2 \times 10^{-6}\ \text{m}^2/\text{s}$, so the penetration distance is 26.8t microns. Experiments [207] show that steam film destabilization occurs in far less than 1 s, so that negligible temperature change occurs at the far end of a simulated 30 μm thickness of melt. Furthermore, during the time step of a shock wave calculation, the thermal penetration in the urania is about 2.8 μm. Accordingly, satisfactory heat diffusion calculations in the molten urania are accomplished with the lattice parameters

$$\delta t = 1\,\mu\text{s}\,; \quad \delta x = 3\ \mu\text{m} \quad \text{for which}\, \alpha\delta t/(\delta x)^2 = 0.22$$

The previously justified uniform pressure in a simulated vapor film renders momentum conservation unnecessary, but mass and energy conservation still require formulation due to interfacial transport. Intuitively or formally from the Rankine-Hugoniot mass conservation equation, vapor particles in contact with the melt have zero velocity. Just one intermediate mesh point between liquid and vapor is

[18] Thermophysical properties of molten urania vary markedly with temperature [245].

recommended in order to ease numerical problems as the film approaches collapse. Now with respect to mass conservation for example, a linear spatial variation of particle velocity V across the thin film is reasonable so

$$V = V_{GB}(z - z_B)/(z_{MB} - z_B) \tag{5.45}$$

where

V_{GB} – interfacial vapor particle velocity

z_B – interfacial position; z_{MB}– melt position (fixed)

The Rankine–Hugoniot mass conservation equation then yields the interfacial velocity as

$$\dot{z}_B = V_{LB} - \upsilon_{LB} G_B \quad \text{with} \quad z_B = \int_0^t \dot{z}_B(\zeta) d\zeta \tag{5.46}$$

and mass conservation for the film as a whole in terms of the mid-point density $\tilde{\rho}_G$ is evidently

$$\frac{d}{dt}[(z_{MB} - z_B)\tilde{\rho}_G] - G_B = 0 \quad \text{with} \quad \upsilon_G = 1/\tilde{\rho}_G \tag{5.47}$$

Reference [233] details all the required finite difference equations, physical processes and the flow chart for a digital simulation. It also demonstrates that in this situation the Knudsen effect [244] does not compromise Fourier's heat-conduction equation. Though physical processes are usually described for both sodium and water, simulations concern only the latter. This bias occurs because steam film destabilization experiments are far more tractable, less expensive and were more immediate to the safety case for Sizewell B.

Cine recordings at AEEW of molten urania poured into water at 0.1 MPa depict an agglomerate of melt and steam descending in the coolant. Prior to a trigger, a state of disequilibrium exists in which latent heat transfer diffuses into the surrounding liquid that is continuously replenished and cooled by the induced turbulence. To replicate something of this situation in a simulation, the temperature of the surrounding mass of water is taken as a uniform 20 °C at 0.1 MPa except for an initially saturated value at the interface. Initial temperatures of a 100 μm thick steam film are assumed saturated throughout, and the melt temperature is taken as 3000 °C. A simulation begins with a 50 μs–5 MPa trigger at the remote end of a 100 mm water column, and the resulting interfacial kinetics and temperatures are shown in Figure 5.8. Prior the shock front's arrival at the interface, a state of quasi-equilibrium appears after initially strong interfacial condensation. Despite the subsequently weaker evaporation and the absence of lateral mass convection in the model, the essentially constant film thickness in this period can be justified by a straightforward energy balance using Table 5.2. After a delay of 66 μs, the shock front arrives at the interface whose displacement to some 8 μm from the melt in 20 μs is largely unresisted. Due to the water column's inertia, small oscillations occur but actual liquid–melt contact is prevented by interfacial evaporation from molecular conduction across the now much thinner film. The absence of a material increase in steam pressure during film collapse is supported by experiments [207] that involve an independent calculation of initial film thickness and the following analysis [206] of its collapse time as a function of trigger pressure.

A simulated trigger pulse appears as a weak shock with an almost (isentropic) sonic propagation speed, so the linear wave equation approximates water slug kinetics by

$$\frac{\partial^2 z}{\partial t^2} = c^2 \frac{\partial^2 z}{\partial x^2} \tag{5.48}$$

where

z – particle displacement
c^2 – $(\text{sonic speed})^2 \triangleq \left.\frac{\partial P}{\partial \rho}\right|_{S\ \text{constant}}$; S – entropy

Under experimental conditions, the Acoustic Impedances[19] of water and steam are respectively

$$Z_L = 1.5 \times 10^6\ \text{kg/m}^2 - s \quad \text{and} \quad Z_G = 2.8 \times 10^2\ \text{kg/m}^2 - s$$

[19] In general $Z = \rho c$.

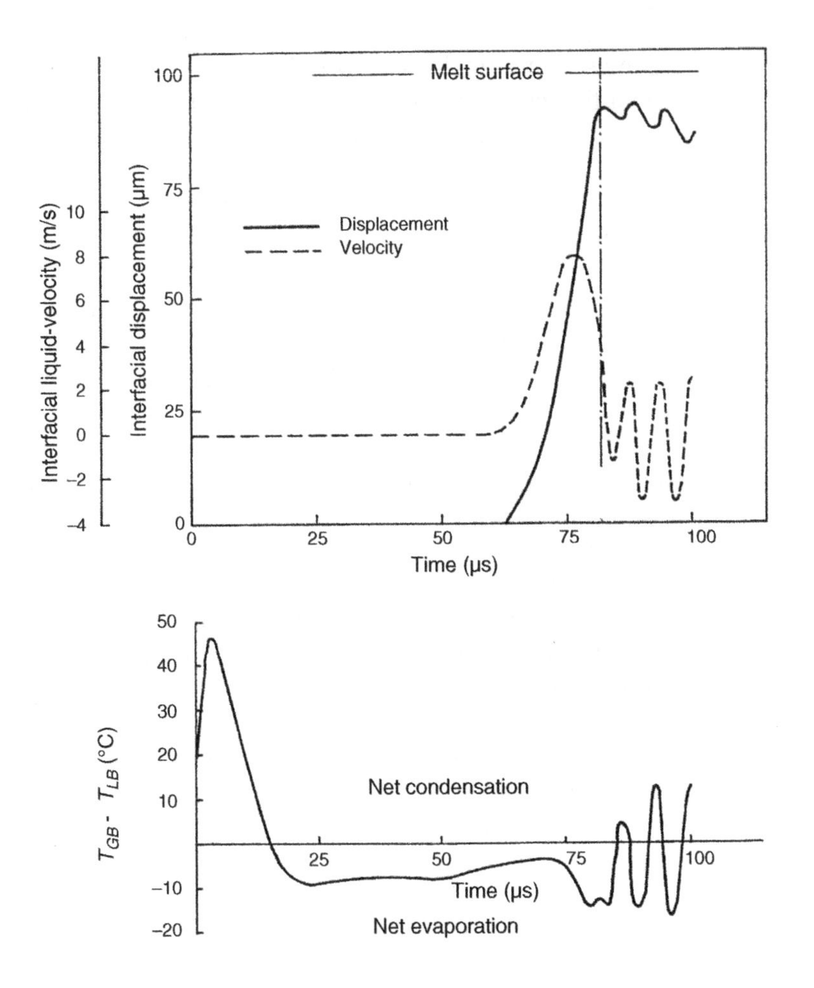

Figure 5.8 Typical Simulation Results for a Melt Temperature of 3000 °C

which yield the interfacial reflection coefficient

$$\Gamma \triangleq (Z_G - Z_L)/(Z_G + Z_L) \simeq -1 \tag{5.49}$$

and this matches the essentially unresisted simulated motion of a liquid–vapor interface. Laplace transform analysis of equation (5.48) for a step function trigger of amplitude P_T and the boundary condition in equation (5.49) gives the transformed interfacial velocity

$$\bar{V}_B = \sum_{n=0}^{\infty} P_T[(2/sZ_L)\exp - 2s\tau]\exp - 2sn\tau$$

where

τ – one sonic transit time along the liquid slug = slug length/c
n – number of forward and backward transits

In the simulated and experimental situations, film collapse occurs after just one transmit time so the relevant interfacial velocity is

$$V_B = 2P_T/Z_L \quad \text{for} \quad t \geq \tau$$
$$= 0 \text{ otherwise}$$

Hence the predicted time for film collapse is

$$T_c = (\text{Initial film thickness})/(2P_T/Z_L) \tag{5.50}$$

Experimental measurements [207] using an electrical probe and an initial film thickness derived from an independent model confirm the above for a solid heat transfer surface below 500 °C.

Computer simulations predict that a steam film over molten urania largely fails to oppose its collapse by a weak shock wave. This prediction is supported by specific steam-film collapse experiments, and MFCI urania tests with sodium or water which are triggered by just the modes of contact. Reactor safety assessments must therefore assume a priori that a contact between molten corium and coolant would produce an MFCI. Moot questions remain concerning the quantity of participating melt and the mechanical energy released (yield). With specific regard to the cited aluminum–water tests in Rig A, Table 5.2 and the discussion of the results in Figure 5.8 suggest that hydrogen generation enhances the thermal conductivity of the steam film, and thereby its stability to the point where a military explosive trigger becomes necessary.

5.7 MELT FRAGMENTATION, HEAT TRANSFER, DEBRIS SIZES, AND MFCI YIELD

With a strong enough trigger and a large enough inertial mass of coolant (>0.1 m), the liquid's final oscillations in Figure 5.8 become large enough for localized contact(s) with the melt. At such instants enormous heat fluxes into the liquid occur. If a relatively low thermal-conductivity blanket were then to re-form, a passive equilibration of melt and coolant would continue. However, the existence of MFCI implies that such vapor blanketing is somehow transiently suppressed, and research into possible mechanisms was initiated by Derewnicki and Hall [243]. Their analysis supported by visualization studies using a platinum wire indicates that acoustic loading (local pressure increases) and Marangoni flows[20] inhibit vapor production during ultra-rapid boiling. It is now suggested that the violent return to thermodynamic equilibrium of locally superheated liquid at contact areas with a melt launches shock waves that create local melt fragmentation [244]. During propagation[21] shock intensity is escalated by further melt fragmentation across its steep frontal pressure gradient, and direct coolant contact is sustained by viscous forces that strip embryonic bubbles from the fragments. Though the above description is in part conjecture the following analysis establishes that the creation of fine debris ($\leq 250\ \mu$m) and a highly efficient heat transfer mechanism are necessary in order to match experimental MFCI time scales.

If an isotropic sphere at a uniform temperature T_M is abruptly immersed in an infinite sea of perfectly stirred coolant, its spatially one-dimensional temperatures thereafter satisfy [224]

$$\frac{\partial T}{\partial t} = \alpha\left(\frac{\partial^2 T}{\partial r^2} + \frac{2}{r}\frac{\partial T}{\partial r}\right) \tag{5.51}$$

with the boundary condition

$$h[T(R,t) - T_L] = -\kappa \left.\frac{\partial T}{\partial r}\right|_R \triangleq \phi(t) \tag{5.52}$$

where

r – radial coordinate; R – radius of sphere

κ – thermal conductivity of the sphere; T_L – coolant temperature

[20] A surface tension gradient in a fluid pulls liquid toward the greater value to create a Marangoni flow.

[21] At around 350 km/h in tin–water experiments [248].

h – external heat transfer coefficient; ϕ – surface heat flux from the sphere

The above partial differential equation has a countably infinite spectrum of eigenvalues, and its Laplace transform solution is the corresponding infinite series [224]. The thermal energy released from the sphere is

$$\text{Surface area} \times \int_0^t \phi(\zeta)d\zeta \tag{5.53}$$

and when the external heat transfer process is highly efficient it is largely represented by the smallest eigenvalue term of the series. Under these conditions the reciprocal of the smallest eigenvalue is called the dominant time constant τ^*, and the heat released per unit mass is approximately

$$\left.\begin{aligned} &E(t) = E_\infty[1 - \exp(-t/\tau*)] \\ &E_\infty = C_p\Delta T; \quad \Delta T = T_M - T_L \\ &\lim_{h\to\infty} \tau^* = R^2/\pi^2 d \end{aligned}\right\} \tag{5.54}$$

where

Because the dominant thermal time constants of square prisms are insignificantly different from those of spheres [224], heat transfer from irregular MFCI debris is taken as that from spheres. Dominant thermal time constants for uranium spheres as a function of external heat transfer coefficient are shown in Figure 5.9 for diameters of 30, 100, 250, and 500 μm. Heat transfer in experimental MFCI with sodium or water is completed within 2–4 μs. Consequently dominant time constants no greater than about 1 μs are involved, and to match these values Figure 5.9 shows that spherical diameters below 250 μm and heat transfer coefficients exceeding 100 kW/m^2 are necessary. Thermal radiation is immaterial because Section 5.4 shows that water is broadly transparent to infrared over these length scales, and even the heat absorbed by a black body from a source at 3500 K corresponds to a coefficient[22] of less than 2.5 kW/m^{2}K.

[22] For a source $T_M \gg T_L; h \simeq \sigma T_M^3$ where σ is the Stefan–Boltzmann constant.

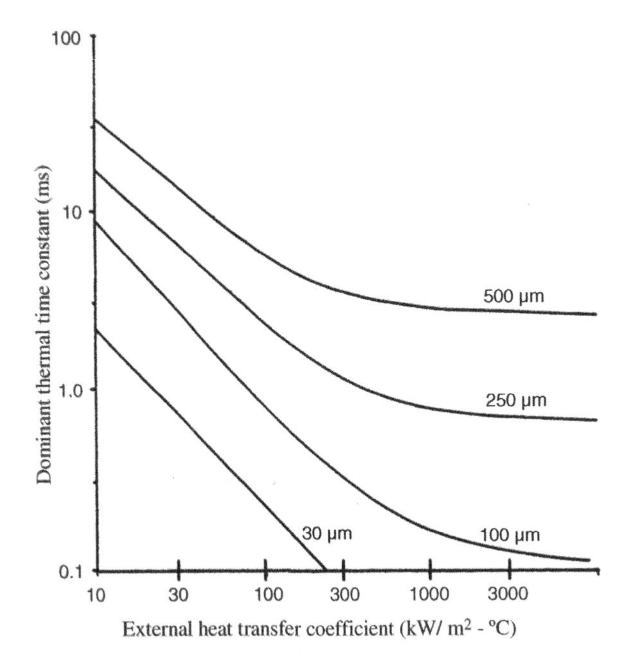

Figure 5.9 The Dominant Thermal Time Constant as a Function of External Heat Transfer Coefficient for Various Diameter Spheres of Urania Having $\alpha = 2.3 \times 10^{-6}\text{m}^2/\text{s}$

After separation from the coolant experimental MFCI debris is graded using a cascaded nest of precision sieves. The size or equivalent diameter d of an individual particle is then taken as the arithmetic mean of the smallest and largest sieve sizes through which it can and then cannot pass. Particle diameters for both water and sodium coolant scan be characterized [246] by a Log Normal probability density function

$$\left.\begin{aligned} &P(d') = \frac{1}{\sigma'\sqrt{2\pi}}\exp\left[-\frac{1}{2}\left(\frac{d'-\mu'}{\sigma'}\right)^2\right] \\ &\text{where} \\ &d' = \log d; \quad \mu' = \mathcal{E}(d') \quad \text{and} \quad (\sigma')^2 = \mathcal{E}(d'-\mu')^2 \end{aligned}\right\} \tag{5.55}$$

However, validation of an MFCI simulation code against experiment is best achieved using the particular a posteriori measured debris sizes. The energetics of an MFCI are clearly influenced by the contact rate of melt fragments and coolant[23] as well as by their sizes. Accordingly if δM of a coarse mixture becomes finely fragmented,($\lesssim 250\ \mu\text{m}$) to $N(\theta)$ particles over the time interval $[\theta, \theta + \delta\theta]$, then assuming spherical debris

$$\delta M = \sum_{k=1}^{N(\theta)} \frac{4}{3} \pi \rho_M r_k^3(\theta) \tag{5.56}$$

where

ρ_M – density of solidified thermite mix
$r_k(\theta)$ – spherical radius corresponding the kth sieve-size at θ[24]

The average of $\{r_k(\theta) | 1 \leq k \leq N(\theta)\}$ is by definition

$$N(\theta) Av.\left[r^3(\theta)\right] = \sum_{k=1}^{N(\theta)} r_k^3(\theta)$$

which substituted into equation (5.56) gives

$$\delta M = \frac{4}{3} \pi \rho_M N(\theta) Av.\left[r^3(\theta)\right] \tag{5.57}$$

If each particle is assumed to liberate its heat independently, then the power released from those newly created during θ to $\theta + \delta\theta$ is similarly derived from equation (5.54) as

$$\delta \dot{Q} = \frac{4}{3} \pi \rho_M N(\theta) Av.\left[r_k^3(\theta) \frac{d}{d\theta} E_k(t - \theta; r_k(\theta))\right] \tag{5.58}$$

where

$E(t - 0; r_k(o))$ = energy released per unit mass of a spherical particle of radius $r_k(o)$ at time $\theta = 0$ as in equation (5.54)

[23] About 60 ms for a complete discharge from the MFTF thermite container.

[24] An unknown, but see later.

Expressing equation (5.57) in terms of the mass creation rate of fine debris W in an interaction

$$W(\theta)\delta\theta = \frac{4}{3}\pi\rho_M N(\theta)Av.\left[r^3(\theta)\right] \tag{5.59}$$

In the limit as $\delta\theta \to 0$, substitution of the above into equation (5.58) yields

$$\frac{d\dot{Q}}{d\theta} = W(\theta)Av.\left\{r_k^3(\theta)\frac{d}{d\theta}E[t-\theta; r_k(\theta)]\right\}\Big/Av.\left[r_k^3(\theta)\right]$$

so

$$\left.\begin{aligned} &\int \frac{d\dot{Q}}{d\theta}d\theta = \int W(\theta)P(t-\theta)d\theta \\ \text{where}\quad &P(\theta) \triangleq Av.\left\{r_k^3(\theta)\frac{d}{d\theta}E[\theta; r_k(\theta)]\right\}\Big/Av.\left[r_k^3(\theta)\right] \end{aligned}\right\} \tag{5.60}$$

If the creation of fine debris occurs at a constant rate W_0 over a prescribed period $\left[0, t_f\right]$ and thereafter is zero, then

$$W(t) = W_0\left[U(t) - U\left(t - t_f\right)\right]$$

where $U(t)$ is the unit step function. By effecting the variable change

$$\theta' = t - \theta \quad d\theta' = -d\theta$$

the convolution (5.60) expands into

$$\left.\begin{aligned} \dot{Q}(t) &= W_0\int_0^t P(\theta)d\theta \quad \text{for } t \leq t_f \\ &= W_0\left[\int_0^t P(\theta)d\theta - \int_0^{t-t_f} P(\theta)d\theta\right] \quad \text{for } t > t_f \end{aligned}\right\} \tag{5.61}$$

which is still intractable unless further assumptions are made. Accordingly, it is conjectured that fine fragmentation by a shock wave is a statistically stationary process [247], so the fraction of each debris size remains statistically constant throughout an interaction. Thus by virtue of the large sample sizes $Av.\left[r_k^3(\theta)\right]$ and $Av.\left[r_k^3(\theta)\frac{d}{d\theta}E[\theta;r(\theta)]\right]$ are largely independent of θ, and can be represented by averages calculated a posteriori from the recovered debris, as

$$\int_0^t P(\theta)d\theta = Av\cdot\left[r_k^3E(t,r_k)\right]/Av\cdot\left[r_k^3\right] \tag{5.62}$$

In the BUBEX simulation of an MFCI test, a one-dimensional look-up table of the above integral (5.62) is first computed separately for each time step using equation (5.54) and the recovered debris spectrum. The mass creation rate of W_0 of fine fragments is estimated from experiment. Alternatively, shock propagation at 100–150 m/s over the small volume of experimental coarse mixtures creates fine debris much faster than their energy release rates, which implies the simpler approximation

$$\dot{Q}(t) = \sum_{n=1}^{S} \delta M_n \frac{d}{dt}E(t,r_n) \tag{5.63}$$

where δM_n is the recovered mass from the nth of S sieve sizes.

The yield of an MFCI is defined as the mechanical work delivered by the expansion of its vapor bubble. Experiments at AEEW measure Yield in terms of the assumed isentropic pressurization of a coolant's argon cover gas, and the efficiency of an MFCI is defined as:

$$\text{MFCI Efficiency} \triangleq \text{Yield/Heat content of participating mass (i.e., debris} \leq 250\,\mu\text{m)} \tag{5.64}$$

However, even if the extrapolation of experimental MFCI efficiency to reactor-scale masses is valid, the resulting Yield alone does not represent the potential damage to a reactor structure. Specifically, the containment vessel of a fast reactor in Figure 5.10 includes primary circuit pumps and intermediate heat exchangers which can focus explosively displaced coolant to exacerbate damage: particularly to the rotating shield. Analytical and corroborative experimental investigations of this phenomenon for a fast reactor are described later in

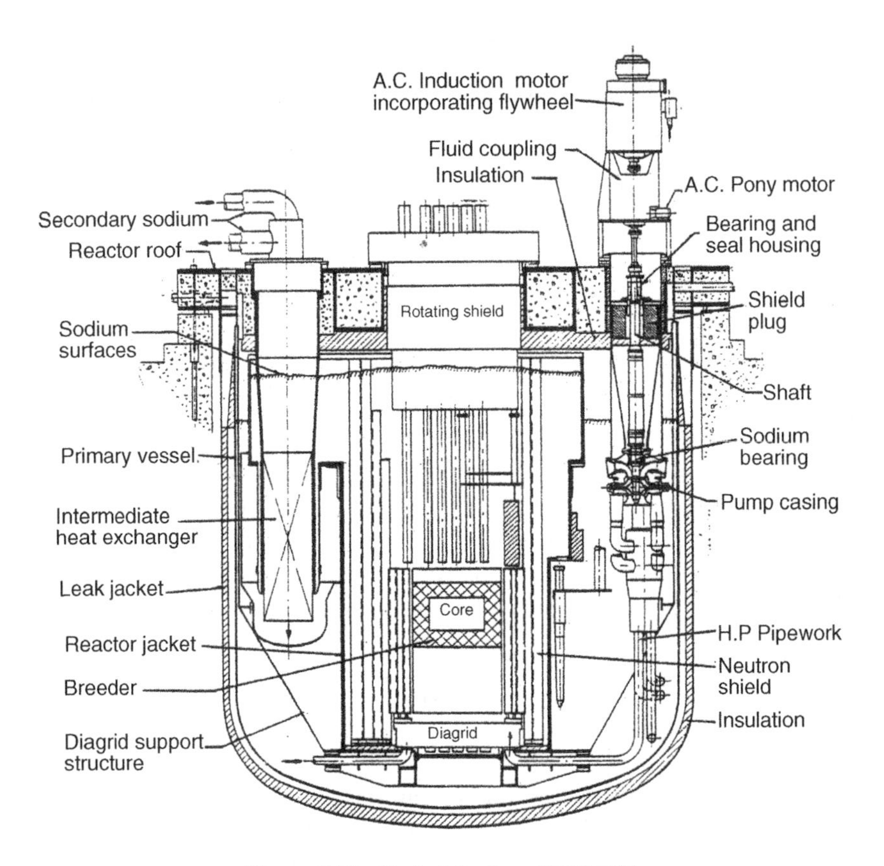

Figure 5.10 Vertical Section of PFR [60]

Section 6.1. In a PWR, the lower core-plate and support casting in Figure 1.4 would constrain the explosion to increase the mechanical loading at the base of its pressure vessel.

To place the explosive violence of an MFCI in perspective, consider a typical Rig A experiment in which 0.5 kg of molten urania thermite at 3500 K reacts with 50 kg of water at 293 K. If passive thermal equilibration were to occur with the whole coolant mass, its temperature would increase to just a modest 299 K. However, the experimental MFCI yield is about 0.16 MJ which corresponds to the kinetic energy of a $1\frac{1}{2}$ tonne vehicle travelling at 55 km/h. Because water reactors and fast reactors have typical fuel inventories of 100 and 20 tonne respectively, the very localized heat transfer in an MFCI appears as potentially catastrophic especially as a 1 GJ yield gives cause for concern with regard to the failure of either reactor vessel. Specifically, granted a Hicks–Menzies isentropic efficiency of 16% and a participating mass of 20% of a fuel inventory,[25] the resulting yields for a molten fuel temperature of 3500 K are

$$\left.\begin{array}{l}\text{Water reactor yield} = 4.8\ \text{GJ}\\ \text{Fast reactor yield} = 1.4\ \text{GJ}\end{array}\right\}\quad C_p\ \text{for}\ UO_2 = 500\ \text{J/kg}$$

Computer simulations involving equation (5.62) are clearly unlikely to precisely replicate experimental measurements of an MFCI yield. Nevertheless by also marrying the condensation mass flux equation (5.27) into apposite fluid dynamics, the BUBEX code in Section 5.8 confirms interfacial condensation as the principal thermodynamic irreversibility that saps material amounts of energy from an MFCI vapor bubble. Consequently, justification is derived for extrapolating the 4 to 5% experimental MFCI efficiency to reactor-scale and thereby significantly enhancing a reactor safety case. Specifically, granted a 4 to 5% efficiency and a molten corium temperature of 3500 K for a Severe Accident in a PWR, then the required participating mass for a 1 GJ event evaluates as about 17 tonne or 17% of the entire fuel inventory. Because experiments indicate that just some 20% of a total melt mass has[26] particle sizes less than 250 μm, a 1 GJ yield corresponds to the non-credible event of the entire PWR fuel inventory in a molten state and in contact with enough water. On the same basis for a fast reactor, but with the highest predicted melt temperature of 5000 K, the required participating mass for a 1 GJ yield is 11 tonne or nearly three times the entire fuel inventory in a molten state! In this context, simulations of Severe Accidents [213,269,270] with distributed neutronics and thermal hydraulics indicate a progressive degradation of a reactor core. Confirmation is provided by the post-accident inspection of the Three Mile Island reactor in which just 8 to 16 tonne of its 100 tonne fuel inventory lay below the lower core-support plate [69]. Furthermore, Section 5.2 outlines the different molten fuel–coolant contact modes as

[25] Forty percent of fine fragmentation is found in some CORECT2 experiments [86], but this is more likely to result from multiple incoherent interactions and/or the debris recovery process from a sodium coolant.

[26] Refer to Section 5.2. Portions of a melt solidify or fail to develop the detonatable morphology.

additional factors that would materially restrict the participating mass. For all these reasons the creation of the participating mass for a single coherent 1 GJ event is considered very improbable especially with present safety systems, operational experience and legislation.

5.8 FEATURES OF THE BUBEX CODE AND AN MFTF SIMULATION

An abrupt release of pressurized Argon from the MFTF charge container and a record of the cover gas pressure transient were made as part of rig commissioning. Various one-dimensional fluid dynamics models proved unsuccessful in calculating this transient despite their use in other MFCI simulations.[27] A spatially higher dimensional model is therefore necessary for authenticity. In fact a sufficiently accurate reproduction of this MFTF commissioning test became the first step in the validation of the MFCI dynamics code BUBEX. Because reactors and MFTF have a fair degree of axial symmetry, two-dimensional spherical coolant dynamics appear promising. In fact the two-dimensional code SEURBNUK had been extensively validated by the earlier WINCON experiments [276] that used scaled models of the fast reactor geometry in Figure 5.10 and contrived low brisance chemical explosives. Accordingly, the comprehensive MFCI model BUBEX was developed to replace SEURBNUK's far simpler representation of a chemical explosion.[28] Salient features of BUBEX are next outlined along with its application to the urania–sodium MFTF experiment in Figure 5.11, which presents just a single interaction.

Inviscid fluid dynamics can be described in general by the Eulerian conservation equations [256]

$$\frac{\partial \rho}{\partial t} + \nabla.\rho \boldsymbol{u} = 0 \quad \text{(mass)} \tag{5.65}$$

$$\frac{\partial \boldsymbol{u}}{\partial t} + (\boldsymbol{u}.\nabla)\boldsymbol{u} = -\frac{1}{\rho}\nabla P + \boldsymbol{g} \quad \text{(momentum)} \tag{5.66}$$

[27] See Refs. [249–254].

[28] Essentially the expansion of a perfect gas.

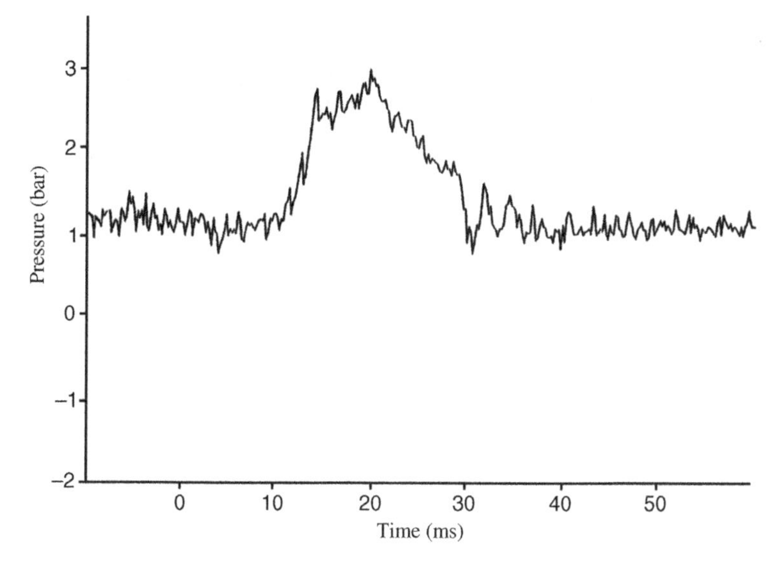

Figure 5.11 Cover Gas Pressure Transient after an MFCI in the SUS01 Urania-Sodium Experiment

$$\frac{\partial \rho I}{\partial t} + \nabla.\rho I \boldsymbol{u} + P\nabla.u = 0 \quad \text{(energy)} \tag{5.67}$$

SEURBNUK solves the above equations using the Mark and Cell Method [257] assuming adiabatic flow and with boundary conditions corresponding to a cover gas space, internal structures and a chemical explosion. As illustrated by Figure 5.12 for the test in Figure 5.11, the interaction of a bubble with internal structures creates an irregular geometry so that calculations of the condensation mass flux and heat transfer to a surrounding coolant would be formidable. However, the simplifying approximation of a spherical bubble having the same instantaneous volume minimizes both the condensing and external heat transfer surfaces and thereby maximizes the computed MFCI yield. Provided that corresponding predictions evolve as small enough, the approximation is sufficient for reactor safety assessments.

Bubble growth in MFTF experiments occurs over some 20 ms, whereas vapor film destabilization in Figure 5-11 occupies just 10 μs. The linearized dynamics of equation (5.17) imply that longer time-scales allow the development of slower and longer wavelength

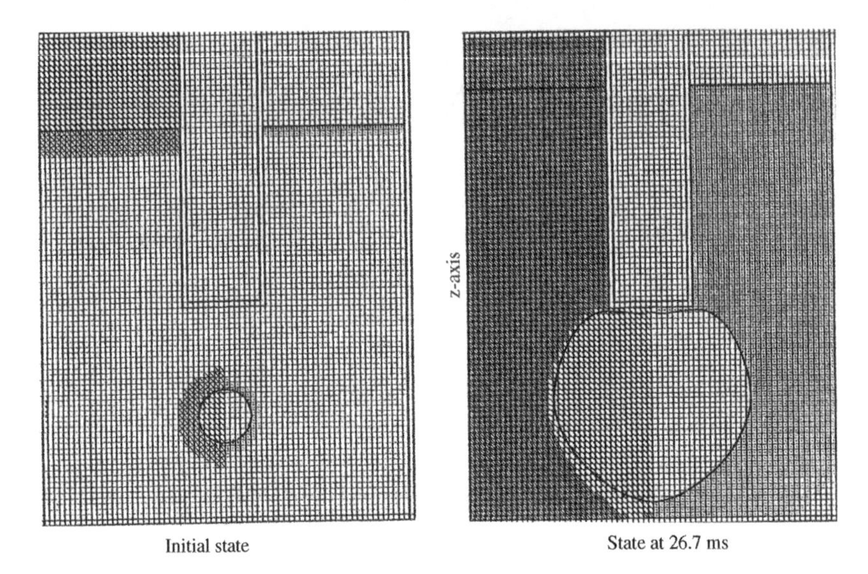

Figure 5.12 SEURBNUK Bubble Geometry in the Urania-Sodium Experiment of Figure 5.11. The Condensation Coefficient is 0.3

Rayleigh–Taylor instabilities for which viscosity is evidently less stifling. Experiments using ethanol–air [258] or liquid–vapor [259–261] systems show that planar interfacial decelerations create highly distended interacting "spikes" that eventually detach. Corradini [254] correlated liquid entrainment into vapor bubbles for scale-model tests at the Stanford Research Institute [264] and Purdue University [265] by

$$\begin{aligned} C_{RT} &= 11.6\rho_L[\sigma_L/(\rho_L - \rho_G)]^{1/4}\ddot{z} \quad \text{for} \quad \ddot{z} \geq 0 \\ &= 0 \quad \text{otherwise} \end{aligned} \tag{5.68}$$

where

σ_L – surface tension of the liquid; $\ddot{z}$ – a largely planar acceleration

However, its accuracy for MFCI simulations is compromised by the one-dimensional planar accelerations and the absence of developmental dynamics. Significantly, a lengthy numerical solution [263] of the non-linear *R–T* equations [262] for a spherical interface between inviscid fluids indicates an entrained mass flux some 5 to 10-times less than equation (5.68) when radial deceleration $\ddot{R}$ replaces $\ddot{z}$. In the absence of a real alternative, *R–T* entrainment mass flux G_{RT} is represented in BUBEX by

$$\tilde{\tau}(t)\dot{G}_{RT} + G_{RT} = C_{RT} \tag{5.69}$$

where C_{RT} is specified by equation (5.68), $\ddot{R}$ replaces $\ddot{z}$,

$$\left.\begin{aligned} \tilde{\tau}(t) &= \mathrm{Min}\{\tau^*(t); 10\} \quad \text{for } \ddot{R} \geq 0 \\ &= 10^{-3} \text{ otherwise} \end{aligned}\right\} \tag{5.70}$$

and $\tau^*(t)$ is defined by equation (5.17). With decreasing deceleration $\tau^*(t)$ becomes ever larger, and to prevent numerical overflow $\tilde{\tau}(t)$ is artificially restricted to 10 s. However, there are no adverse consequences as the durations of MFTF transients are markedly shorter ($< 1/10s$). Once an interfacial liquid accelerates into its vapor, the interface restablizes but no information on this behavior appears available. Under these conditions, BUBEX arbitrarily assigns the time constant of 1 ms. The above uncertainties dictate that MFCI simulations should be scoped with 5 to 10 factor scalings of the above mss flux G_{RT}. In this respect, Severe Accident calculations are only required to be conservative rather than to meet the $\pm 10\%$ accuracy for engineering design.

Liquid entrainment into an MFCI bubble also occurs as the interface is scoured by turbulent vapor [236] or as it brushes around damaged internal structures [71]. There is a paucity of data on entrained droplet sizes, but aerosol experts [171,266] suggest a range of 1 to 100 μm. By virtue of their large surface area to mass ratio, entrained droplets are potentially efficient removers of released fission products by dissolution or adhesion [104,171,266,267]. Asymptotic values for the dominant thermal time constants of the suggested size range are obtained from equation (5.54) as

$$\left.\begin{aligned} &0.15\,\mu\mathrm{s} \leq \tau \leq 1s \quad \text{for water} \\ &0.53\,\mu\mathrm{s} \leq \tau \leq 3.6\,ms \quad \text{for sodium} \end{aligned}\right\} \tag{5.71}$$

so smaller droplets are likely to be vaporized within a turbulent bubble. Their radioactive burdens would then be precipitated as larger

Table 5.5
Ratio of Expectations in Equation (5.72) for $\alpha \ll \beta$

Probability Density	Falling Triangular	Symmetric Triangular	Rising Triangular	Uniform
$\varepsilon(D^2)/\varepsilon(D^3)$	$1.67/\beta$	$1.56/\beta$	$1.25/\beta$	$1.33/\beta$

agglomerates. Those larger than about 10 μm would rapidly settle-out under gravity to be trapped in the turbulent coolant. The expected area of a mass M_D of entrained droplets is readily derived as[29]

$$\varepsilon(A) = 6M_D\rho_L\left[\varepsilon(D^2)/\varepsilon(D^3)\right] \quad (5.72)$$

where

D – a droplet diameter; ε – statistical expectation (mean)

If the probability density function of D has lower and upper bounds α and β with $\alpha \ll \beta$, the above ratio of expectations for various rudimentary distributions closely approximates those in Table 5.5. Granted a wide size-spectrum the area available for aerosol scrubbing is seen to be largely dictated by the largest droplets: and not by the population's detailed statistics. This fact should simplify experiments to provide a sounder basis for aerosol scrubbing simulations like those in BERTA [25] and BUBEX.[30] The FAUST experiments [267] concerned liquid entrainment by permanent gas bubbles, so that droplet longevity was not foreshortened by evaporation or encounters with fuel fragments. Accordingly the observed highly efficient aerosol scrubbing process in FAUST might not be replicated in a reactor situation. Realistic lifetimes and heat-transfer data for entrained droplets are clearly essential pre-requisites for specifying the radiological source term in a safety assessment. In the absence of such data, BUBEX preferentially evaporates Rayleigh–Taylor droplets before those of the surrounding coolant, so as to provide a conservative assessment.

[29] The ratio $\varepsilon(D^3)/\varepsilon(D^2)$ is often termed the Sauter diameter [308].

[30] The embedded BURST subroutine was developed by Dr. Kier at UKAEA, Culcheth.

Experiments establish that the collapse rate of steam bubbles is reduced by just 10% with the presence of a 15% molar concentration of permanent gas. With the inhibiting effect of permanent gases on power station condensers[31] borne in mind [219], researchers [236,267] have concluded that violent turbulence must exist within an MFCI bubble. Also it is reasonable to conjecture that similar turbulence develops in its surrounding liquid, but an apposite correlation for heat transfer from the liquid interface was not available during BUBEX development. Previously cited MFCI models adopt quite speculative heat transfer relationships or none. However, in order to discard Hicks-Menzies efficiencies in favor of the some six times smaller 4 to 5% experimental values for reactor safety assessments, a patently conservative representation of heat transfer from a liquid interface is required. If this heat transfer process were to be inefficient, then a weakened condensation mass flux would be less effective in sapping energy from a simulated MFCI bubble.

Turbulent fluid flow around a body results in an attached laminar flowing boundary layer whose periphery is scoured by eddies induced by viscous shear [219,268]. Steady-state heat transfer is then often represented by molecular conduction across the boundary layer augmented by a dynamic diffusion process associated with the eddies. Formally

$$\phi = -\left(\kappa + \rho\, C_p E_H\right)\frac{dT}{dy} \tag{5.73}$$

where beside the usual nomenclature,

$E_H \triangleq$ Eddy diffusivity of heat

y – perpendicular distance outwards from the body

Due to their low molecular conductivities, steady-state heat transfer to turbulent water or steam is totally dominated by eddy diffusivity so experimental correlations [64,117,143,219] involve only Reynolds Number terms. On the other hand, correlations for highly conductive liquid metals involve a sum of κ and E_H terms [64,117]. To provide conservative predictions of MFCI yield, BUBEX models heat transfer

[31] Hence the use of deaerators in boiler feed trains.

Table 5.6
Computed Cover Gas Parameters for the Test in Figure 5.11

Condensation coefficient (σ)	0.0	0.1	0.2	0.3	0.4
Peak cover gas pressure (bar)	82	9.3	3.2	1.9	1.5
Rise time (ms)	14	25	30	30	27
Experiment 2.7 bar after 19 ms					

from a liquid sodium interface as a moving boundary molecular diffusion process like that in Section 5.6. The thermal penetration distance, and therefore lattice length for the 30 ms transient in Figure 5.11 evaluates from equation (5.44) as around 3.2 mm.[32] To preserve a conservative calculation and feature the transiently developing boundary layer, the "shell" of surrounding liquid sodium correspondingly thins with bubble expansion. Note that this approximation is evidently inapplicable to urania–water interactions for which a bona fide correlation for heat transfer from the liquid interface remains as necessary data.

Figure 5.12 illustrates a simulation of experiment SUS01 and the influence of MFTF structure on bubble dynamics. Because the fluid dynamics computation fails if the bubble actually touches structure, considerable skill[33] is necessary to restart the simulation just before this event with a large enough yet close enough configuration. Due to the unknown purity of the sodium, simulations of the test in Figure 5.11 were made with the range of condensation coefficients in Table 5.6. Because the cover gas pressurization is seen to reduce markedly with increasing values of σ, the adopted conservative model of heat transfer from the liquid interface does not materially restrict energy dissipation. Consequently, it is sufficiently sound for present purposes. In order to provide the closest match to the 19 ms rise time in experiment SUS01, the condensation coefficient for MFTF's sodium contents was selected as

$$\sigma = 0.1 \tag{5.74}$$

[32] The transformation $T' = rT$ maps a spherical into the planar case.

[33] By Mark Turrel.

Table 5.7
Comparison of Bubble Expansion Parameters

	Experiment	BUBEX	Lossless
Peak cover gas pressure (bar)	2.7	9.3	82
Work done on cover gas (kJ)	24	72	243

Table 5.7 compares the peak cover-gas pressurizations and corresponding work done on the cover gas for experiment SUS01 with this BUBEX calculation and a lossless bubble expansion ($\sigma = 0.0$). The BUBEX simulation actually accounts for 78% of the energy dissipation relative to the lossless case. It is concluded that Hicks–Menzies efficiencies are indeed over-predictions and that sound reasons exist for using the some 6 times lower 4 to 5% experimental values at reactor scale.

CHAPTER 6

Primary Containment Integrity and Impact Studies

6.1 PRIMARY CONTAINMENT INTEGRITY

Safety assessments for nuclear plants include the effects of Severe Accidents[1] on the integrity of the reactor vessel (the primary containment). Water and fast reactor vessels are potentially subject to short time-scale pressure and fluid-impact loadings from MFCI. In addition, fast reactor vessels might suffer slower pressure-induced loadings over several seconds due to conventional vaporization by larger sized corium debris, and in the United Kingdom these are termed Q^*-events [202]. French and UK fast reactor designs are the pool type shown in Figure 5.10, in which a double-skinned reactor vessel houses intermediate heat exchangers, pumps and access areas to core components. Their typical 2000 tonne sodium inventories provide enormous heat sinks, and if 50 tonne of molten corium at 5000 K were to passively equilibrate, the bulk sodium temperature would rise innocuously from 900 K at 1 bar to less than 950 K. The alternative loop-type fast reactor

[1] For fast reactors, Severe Accidents are more usually termed hypothetical core disruptive accidents (HCDA).

Nuclear Electric Power: Safety, Operation, and Control Aspects, First Edition.
J. Brian Knowles.
© 2014 John Wiley & Sons, Inc. Published 2014 by John Wiley & Sons, Inc.

design, like the German SNR300, has intermediate heat exchangers and pumps outside the reactor vessel, so the primary circuit layout is not dissimilar to that of a PWR. For obvious safety reasons code validation experiments have generally[2] used water rather than molten sodium as the coolant. If it can be demonstrated that primary circuit elements can withstand the spectrum of Severe Accident loadings, then the surrounding reinforced concrete building (the secondary containment) would provide a second, but redundant, barrier between radioactive fission products and the local population. Descriptions follow of scaled fast reactor experiments aimed at validating calculations of transient stresses in primary circuit components that would occur from MFCIs. These investigations establish the soundness of the structural analysis codes used in safety assessments for both water and fast reactor designs. Later sections outline experimental and theoretical research into the robustness of reinforced concrete containments [106,275] to potential impacts of airborne plant fragments ("missiles") and aircraft.

Operating temperatures of around 500 °C and crowded interiors complicate replica scaling of a reactor's interior. Models used in experiments have sought to represent isolated features (e.g., hemispherical dome on a cylinder [273]) or complex 1/20th scaled versions of internal structures [274,276]. The COVA series of experiments [88] on pool and loop fast reactor designs were an international collaboration between AEEW, AWRE and the European Joint Research Center Ispra. Initial results were used to validate the two-dimensional axisymmetric fluid dynamics codes ASTARTE (Lagrangian) and SEURBNUK (Eulerian). Because gross fluid motions around internal structures are more readily represented by a fixed-geometry Eulerian mesh, the latter became the focus of research activity. Structural loadings from SUERBNUK formed inputs to the mathematically decoupled structural dynamics code EURDYN, which at the time had just one-dimensional modules. Later code developments created two- and three-dimensional subroutines for analyzing the WINCON [276], STROVA and reinforced concrete[3] tests.

The rapid structural loadings from MFCIs increase the yield stress of steels by some 25% as illustrated by Figure 6.1 Because EURDYN modules did not allow for this strain-rate enhancement, scoping

[2] An exceptional use of sodium is described in Ref. [271].

[3] A three-dimensional calculation is crucial for a reinforced concrete impact [275].

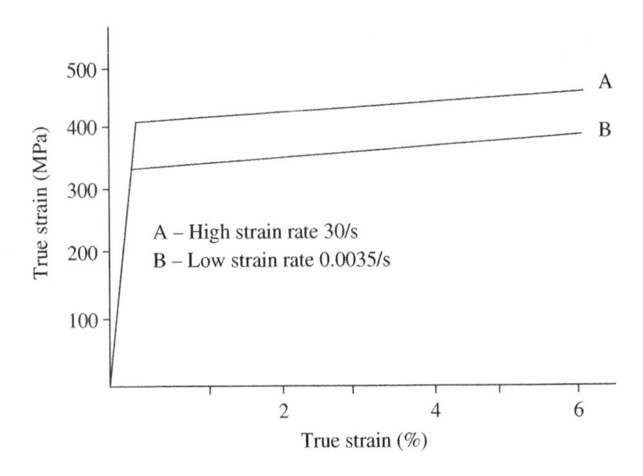

Figure 6.1 Strain Rate Enhancement for Ferritic Steel

calculations were necessary to span the experimental measurements by using low- and high- strain rate data. Repeatability of a load inducing detonation is patently crucial for code validation, but evidently lacking in experimental MFCIs. Accordingly chemical explosives have been used as MFCI simulants. At the Enrico Fermi Plant a TNT charge was actually detonated in a sodium-filled vessel [271], but Figure 6.2 reveals that the energy release transient from a high explosive differs materially from an MFCI event. A detonation is characterized by the shock wave spatially leading the place of energy release [203],whose release rate for an MFCI is restricted by heat diffusion within the resulting fine debris. On the other hand the far faster energy release and shock intensity from a military explosive result from a virtually concomitant rupturing of chemical bonds at the shock front. It follows that a closer match to an MFCI detonation necessitates a much smaller shock speed than in a chemical explosive. By coating easily compressed polystyrene granules with pentaerythritol tetranitrate (PETN) and then expanding the mixture in a mould, shock propagation speed is markedly reduced[4] to achieve the closer match shown in Figure 6.2. Nevertheless the low-density explosive (LDE) still releases some 20% of its energy as an over-sharp shock wave. To obtain a validation of EURDYN's finite element modules under conditions closer to those of an MFCI, the STROVA rig [273] with a vacuum gun was used. In essence this gun consists of an

[4] Visualize shock propagation in terms of a cascade of lossy spring-linked point masses.

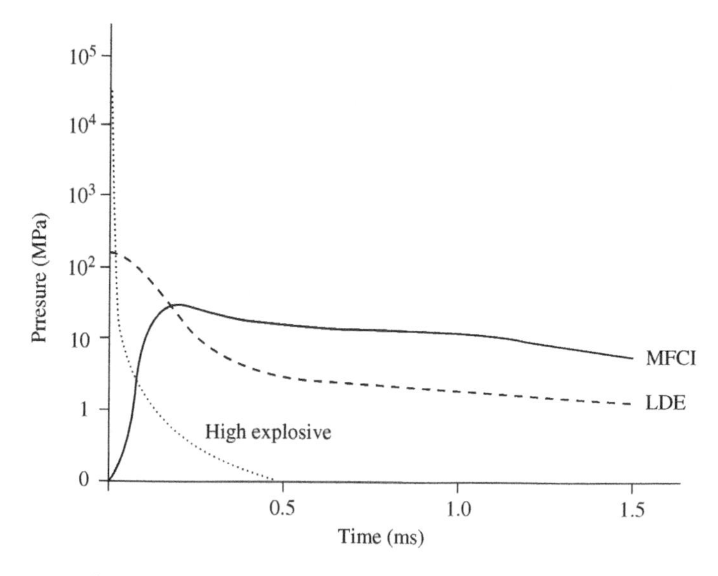

Figure 6.2 Typical Characteristics of Detonation Processes

evacuated barrel with a diaphragm seal at each end. A triggered breach of the upper diaphragm drives a weight downwards[5] under atmospheric pressure so as to rupture the lower diaphragm, which then drives a piston whose impact on a hydraulic fluid loads a selected test piece.

The internationally sponsored COVA experiments [88] progressively added different internal components in $1/20^{th}$ scale models of loop- and pool-type reactors until all the major axisymmetric features became broadly represented. An LDE charge was detonated on the axis of symmetry and each test was repeated at the previously cited research centers to eliminate systematic errors. Deforming structures were represented by thin-shell segments within SEURBNUK, and more complex or "rigid" structures via the link to EURDYN-1. Experimental pressure loadings below the liquid surface and strain patterns were generally well-represented by the two codes. However, impact loads on a model roof and the actual magnitude of strains did not meet the desired accuracy of $\pm 20\%$. Because the COVA program considered just axisymmetric structural components, discrete three-dimensional resistances to fluid flow had to be "smeared". Later WINCON [276] tests

[5] At speeds as high as 100 m/s (216 mph).

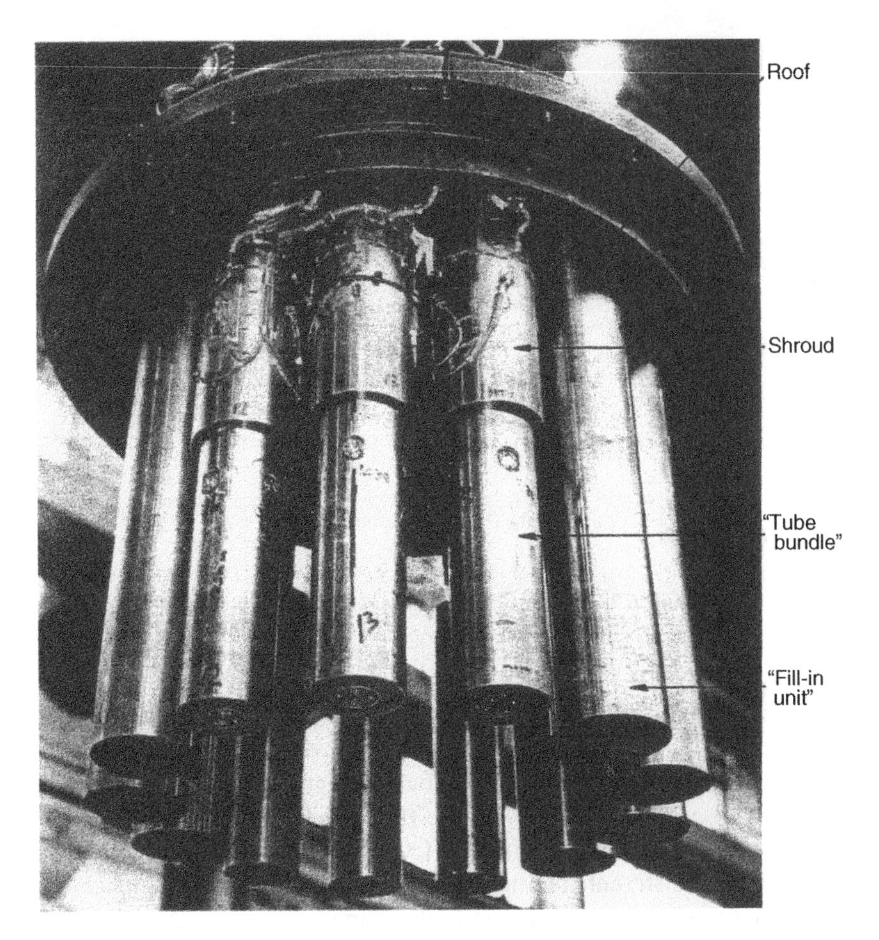

Figure 6.3 WINCON-15: View of the Ring of IHXs and Pumps Attached to the Roof of a 1/20th Scale Model of a Pool-Type Fast Reactor

involved more realistic models of the internal structures within a pool-type fast reactor as depicted in Figure 6.3 As in the COVA series the number of different components was increased progressively through the series as illustrated by Figure 6.4, but appropriate stress–strain calculations were now performed by three-dimensional EURDYN-3 modules. Significant asymmetries in fluid flows or vessel loadings were discovered not to have been introduced by the more realistic arrangements. The rotating roof shield evolved as the weakest component in an early CDFR design, and impact by a moving mass of sodium from a large enough MFCI could raise it sufficiently to allow an escape of

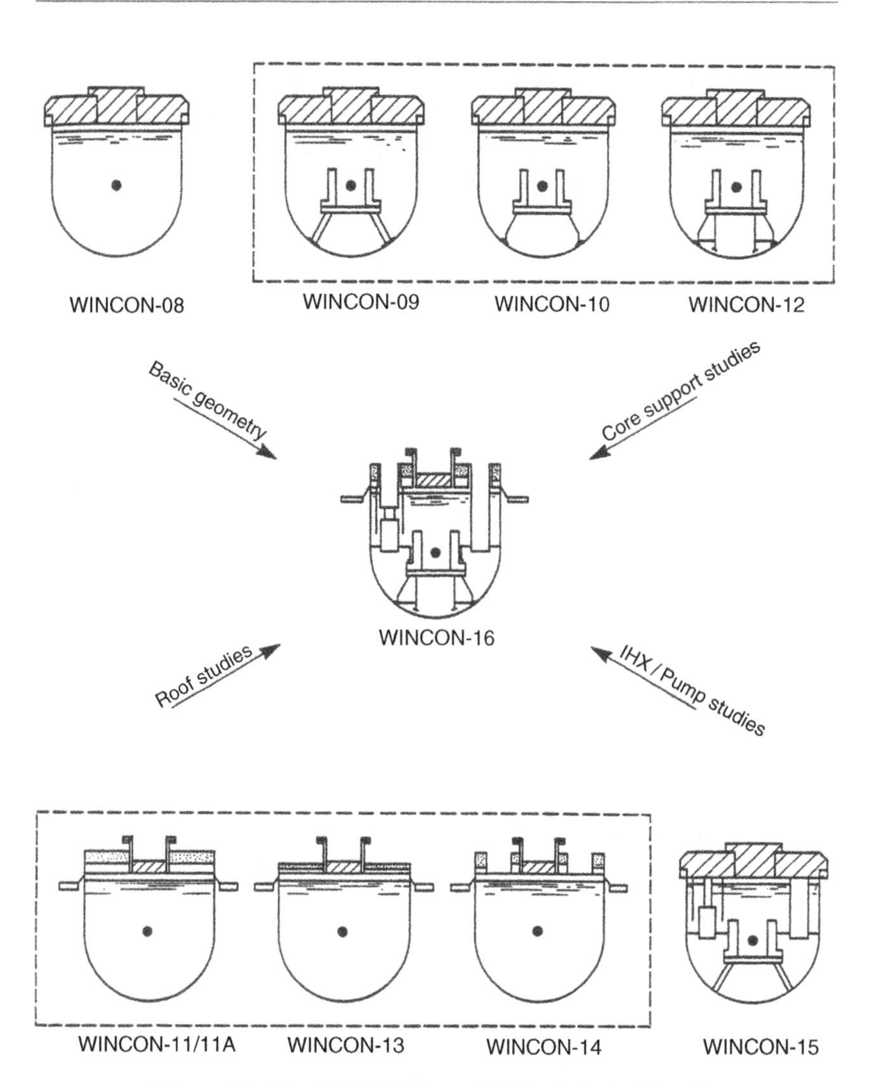

Figure 6.4 The WINCON Tests with Hemispherical Vessels

sodium and corium into the reactor hall. Accordingly, the effectiveness of dip plates, deflector plates and crushable shields were investigated as means of protecting the roof. Though crushable material reduced peak loadings by more than a factor of 4, design difficulties settled the adoption of a simple dip plate that provided a predicted 50% reduction. With a refined version of SEURBNUK [276], sodium impacts on the model roof were found to be transformed into a series of weaker pulses by successive recompressions of the coolant. This dynamic coupling between fluid and structural dynamics also reduced roof stresses by about 50%. As well as motivating design developments, it shows that the decoupling of SEURBNUK and EURDYN calculations is inappropriate. Ref. [276] provides a detailed account and critique of the WINCON experiments.

A more stringent validation of EURDYN modules was undertaken by the STROVA rig studies [273] on two basic types of reactor components. One set represented scaled roof elements like circular and annular plates and then progressed to a composite representation of the whole CDFR roof. The other set concerned hemispheres on cylinders which characterized portions of the fast reactor vessel itself (and clearly that of a water reactor as well). An initial test program employed aluminum models which being largely independent of strain-rate enhancement provided a more incisive test of the code. Then with ferritic steel experiments, the plate and annular models were deflected by a few thicknesses, while the composite roof was distorted by around one-quarter of its depth to take the metal into its plastic regime. Hoop strains of up to 2% were sustained by the hemisphere on cylinder models, and by repeat experiments an estimated accuracy of ±5% was achieved. A similar accuracy for other strain measurements was obtained from recordings at points of symmetry. Pressures applied to specimens were taken by tourmaline transducers to an accuracy of ±1.5%.

Strain rates up to 5/s were measured during tests on plain and annular plates, and EURDYN-1 calculations with low and high strain-rate data predicted the observed maximum deflection to within +3% and −21% respectively. By imposing constraints to match boundary and symmetry conditions, a 45° sector of the composite roof model was sufficient for three-dimensional calculations. Because experimental strain-rates of up to 25/s were observed in the lower roof plate, calculations with the higher strain-rate data generally gave the better match. However, the predicted stress transients were too quick and the final 8 mm deflection of its inner edge was underestimated by up to 25%. Refinement of metallurgical data would clearly have enhanced predictive accuracy. Moreover, "dimpling" of this test specimen again indicates a significant interaction between fluid and structural dynamics. Consequently SEURBNUK and EURDYN are required to be mathematically coupled, but foreclosure of the European fast-reactor project forestalled the necessary developments.

6.2 THE PI-THEOREM, SCALE MODELS, AND REPLICAS

Studies of the impacts between missiles and structures have been a long and continuing military activity. An early example is the evolution of square-cornered Norman castles into the rounded structures of Edward I so as to better resist the impact of large catapulted rocks. By the sixteenth century mathematics and chemical explosives had enabled the embryonic formation of modern-style artillery units with development focussed on high-velocity kilogram-size ordnance for effective mobile deployment. During the English Civil War (1642–1651) success revealed itself in the form of 10 kg cast-iron cannon balls with sufficient kinetic energy to reduce stone-built castles to ruins. During operation "Desert Storm" starting January 16, 1991 US tanks fired projectiles of some 9 kg with supersonic muzzle velocities as great as 1900 m/s. Though the rotating machinery and pressurized components in nuclear power plants can produce potentially damaging missiles, their masses and velocities are radically different from the military. For instance a turbine failure at Calder Hall in 1958 created a number of subsonic missiles of order 1 tonne [278]. The probability of plant failures producing missiles with damage potential has been estimated as 10^{-4} to 10^{-5} per operating year [279]. Impacts on reactor structures from subsonic external sources such as crashing aircraft are also probable, and that for a heavy fighter (e.g.,Tornado) is judged to be about 10^{-6} per year. Though light aircraft pose virtually no hazard to reactor containments, they can potentially damage fuel stores or control rooms with the same probability of 10^{-6} per year. Large airliners are considered to have an impact probability of at least one order less than 10^{-6} per year. These power plant impacts produce far less local heating than do military projectiles, so that material properties like creep strength are far less adversely affected. Military data are therefore inappropriate for reactor safety assessments for which the relevant UK studies began in earnest [106] around 1977. In the context of a nuclear power plant, a missile is described as soft if a significant fraction of its deformation is orders of magnitude greater than that of the target. Missiles from disintegrating power plant items generally suffer a dissimilar deformation to their target, and are designated as hard. Table 6.1 summarizes the pertinent parameters of these radically different non-military type impacts.

Table 6.1
Potential Missile Hazards to Reactor Plant [106,290]

Missile Category	Example	Mass (tonne)	Velocity (m/s)
Soft	Military aircraft	20–50	150–300
	Civil light aircraft	1–25	60–90
	Boeing airliner 707	100–320	100
	Steam-drum end	25	80
Semi-hard	Pipe-line end cap	0.03	170
Hard	Valve stem or bonnet	0.15	150
	Turbine disc fragment	$\lesssim 1.5$	150

Because experiments with full-size missiles and nuclear plant structures are impractical, scale models are a necessity. Appropriate scaling rules can be developed either from the fundamental equations or by the presently more convenient route of dimensional analysis [280]. The essence of dimensional analysis is the Buckingham Pi-Theorem, which characterizes a physical process in terms of the minimum number of dimensionless combinations of its pertinent variables. If F denotes a finite polynomial in the variables $\{x_1, x_2, \ldots . x_n\}$, then F is homogeneous of order integer m if and only if

$$F(x_1, x_2, \ldots . x_n) = \sum_j a_j \left(\prod_{p=1}^{n} x_p^{k_{jp}} \right) \tag{6.1}$$

where for all integer j

$$\sum_{p=1}^{n} k_{jp} = m \quad \text{and} \quad a_j \in \mathbb{R}$$

Euler proved that the most general solution of

$$\sum_{p=1}^{n} x_p \frac{\partial F}{\partial x_p} = 0 \quad \text{is} \quad F(x_1, x_2 \ldots . x_n) = 0 \tag{6.2}$$

Table 6.2
Parameters of a Low-Velocity Missile Impact

Variable	Symbol	Dimensions
Missile diameter	d	L
Missile length	h	L
Nose radius of missile	r	L
Angle of nose	α	–
Missile density	ρ_m	ML^{-3}
Yield stress of missile	σ	$ML^{-1}T^{-2}$
Missile velocity	V	LT^{-3}
Angle of impact	β	–
Target thickness	L	L
Target width	w	L
Target density	ρ_t	ML^{-3}
Yield stress of target	S	$ML^{-1}T^{-2}$
Strain	ε	–

where F is a homogeneous function. Based on the above, Buckingham's Pi-Theorem asserts that

> *"The physical measure P_1 of a quantity which depends only on other quantities with measures $P_2, P_3, \ldots P_n$ reduces to one with the minimum number* m *of their non-dimensional combinations."*

Its interpretation in physical terms is that two realizations of a process with the same complete set of Pi-terms have the same dynamic behavior.

Though a systematic procedure [280] exists for deriving the dimensionless Pi-terms of a physical process, those describing the interaction between a missile and a structure can be determined by physical insight when there is a negligible temperature rise in the latter from its plastic deformation. Table 6.2 lists the determining variables under this condition, and the pertinent dimensionless terms can be deduced by adopting missile diameter, target density and target strength as the reference parameters. These can then be combined with impact velocity to provide the additional Pi-terms to characterize the target response as

$$F = \left(V\sqrt{\rho_t/S}; h/d; L/d; r/d; w/d; \alpha; \beta; \rho_m/\rho_t; \sigma/S, \varepsilon\right) = 0 \quad (6.3)$$

in which the constants and exponents of the Pi-terms are to be determined experimentally.

If the dynamics of a scale model are to replicate its prototype, then a constant scaling of the geometric lengths alone would be inappropriate in the present context. Specifically suppose the geometric lengths in equation (6.3) are scaled by λ, and to assist visualization model strains are to match those of the actual structure. Arbitrary scaling of stresses and densities by say ϕ and μ then necessitates a functionally dependent scaling of the model velocity to achieve the same Pi-term. Using primed variables for the model, it is therefore required that

$$V'\sqrt{\rho'_t/S'} = V\sqrt{\rho_t/S}$$

so the model velocity for dynamic similarity must be scaled according to

$$V' = V\sqrt{\phi/\mu} \quad \text{with} \quad \phi \triangleq \rho_t/\rho'_t \quad \text{and} \quad \mu \triangleq S/S' \qquad (6.4)$$

By definition, a replica model[6] has scaled variables that reproduce the set of all dynamically characterizing Pi-terms of the prototype.

Early international experiments [68,106] to validate replica scaling techniques for the study of missile–concrete impacts involved microconcrete with an appropriately scaled aggregate mix and steel reinforcement mesh to represent a prototype. Dynamically similar tests[7] at AEEW employed the three mass-sized pairs in Table 6.3, and for each pair three different bonding reinforcement quantities of $^1/_8$, $^1/_4$ and $^1/_2$% EWEF[8] were investigated for the concrete panels. Visually identical overall damage patterns were produced for each different reinforcement, and the excellent consistency of the measured target penetration velocities is shown in Figure 6.5. These tests adopted the replica scaling in Table 6.4 so that the reinforcement has identical strength, yield and elastic modulus as the prototype. Also the microconcrete target is manufactured to provide the same compressive and

[6] Different materials from a prototype are sometimes used: see Ref. [297], for example.

[7] Same Pi-terms.

[8] Percent EWEF – % of a cross-sectional area occupied by the same square (EW) steel mesh reinforcement just below the surface of each panel face.

Table 6.3
Hard Missile–Target Combinations in Replica Scaling Studies

Missile		Target	
Diameter (mm)	Mass (kg)	Diameter (m)	Thickness (mm)
313	490	6.0	640
120	27	2.3	246
40	1	0.767	82

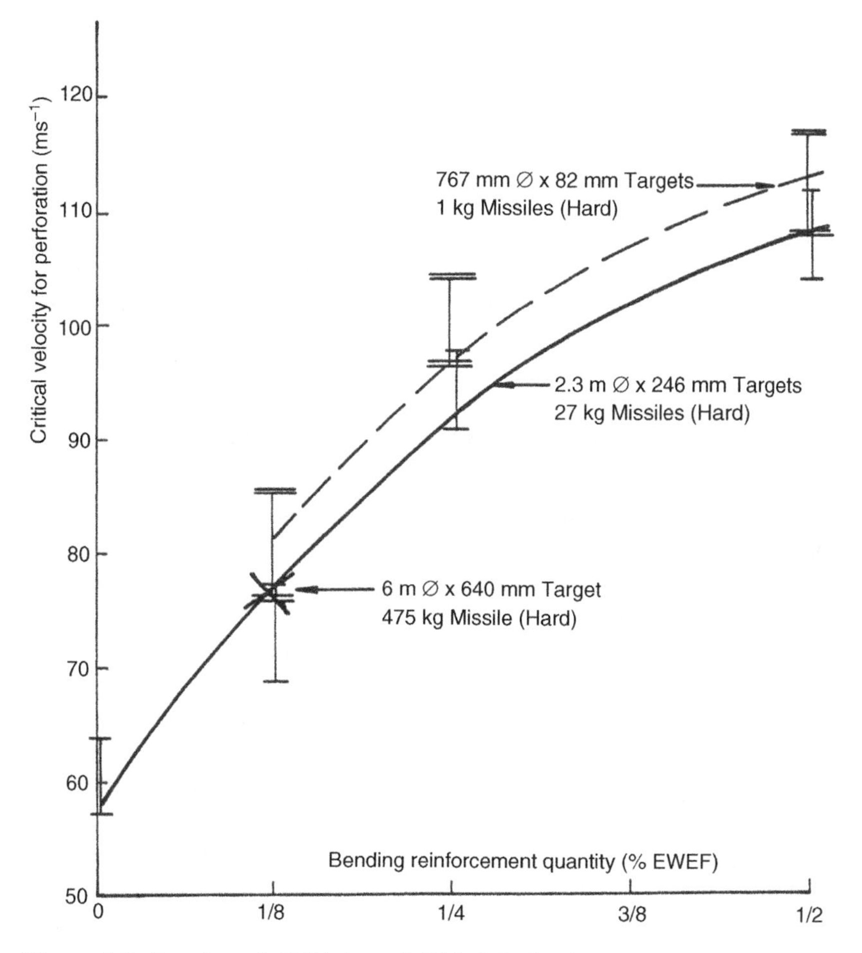

Figure 6.5 Experimental Validation of Critical Perforation Velocity with Bending Reinforcement Quantity for Three Sizes of Concrete Target

Table 6.4
A Consistent Set of Replica Scale Factors

Variable	Length	Velocity	Density	Stress	Strain
Scale Factor	λ	1	1	1	1

Variable	Mass	Time	Force	Frequency	Strain-Rate
Scale Factor	λ^3	λ	λ^2	$1/\lambda$	$1/\lambda$

tensile strength as a typical constructional concrete. Because crack widths and spacings in a concrete structure's flexural response markedly depend [281] on the bonding strength between the concrete and its steel reinforcement, the production of such carefully scaled microconcrete required a dedicated laboratory facility. Actual impacts on a reactor containment induce strain-rates in the range 0.01 to 1.0/s, so that the increased dynamic strength of a replica's steel reinforcement becomes a major difficulty should it become much smaller than the prototype [283,284]. Figure 6.6 illustrates the variation in dynamic strengths of reinforced concrete materials at high straining rates.

6.3 EXPERIMENTAL IMPACT FACILITIES

The following experimental impact facilities at AEEW were typical of those involved in collaborative agreements with French and German companies. As explained in Section 6.2 the provision of micro-concrete which accurately replicates the bonding between actual concrete, aggregate and steel armatures is of paramount importance. Because commercial suppliers were considered unable to meet the required standards of consistency, a small manufacturing laboratory was constructed with an associated suite of measuring devices to test the starting materials of cured concrete and reinforcing steel [106]. Considerable care was also taken to ensure the precise locations of reinforcements. Some 70 test specimens were taken in the form of cubes, cylinders and beams from each concrete mix, and some 10 to 20 pull-out discs were included in each target to assay the quality of the cured material. These tests generally confirmed a ratio of compressive to tensile strength of 10:1, which is typical of a prototype. Material data was

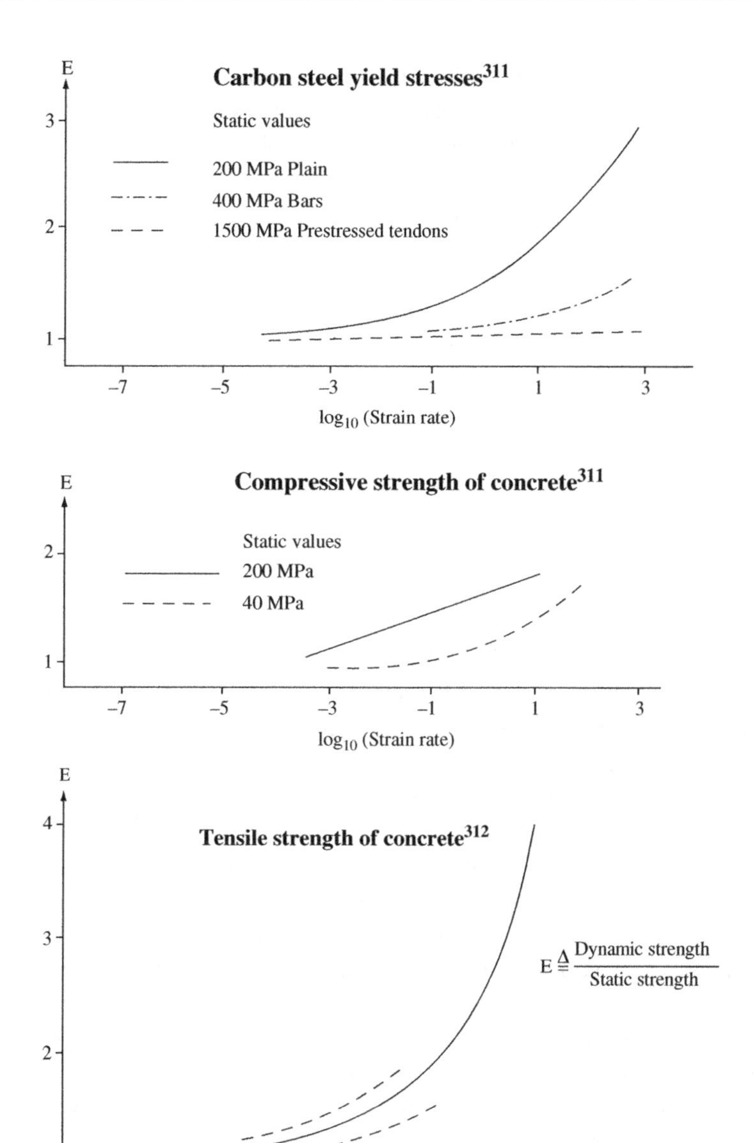

Figure 6.6 Typical Strain-Rate Enhancements for Reinforced Concrete

obtained using a 3 MN hydraulic press which could induce constant, ramp, triangular, sinusoidal or step loadings on a variety of test geometries.

To satisfy the velocities for replica scaling shown in Table 6.4, the two compressed-air guns shown in Figure 6.7 were constructed. The earlier and smaller Missile Launcher had the performance specifications:

Maximum projectable energy	3 MJ
Interchangeable barrel IDs	50–300 mm
Projectile velocity range	10–350 m/s
Target abutment	$\simeq$ 40 tonne

Compressed air within a reservoir provided the variable energy source and thereby a variable projectile velocity. A tubular barrel was separated from the reservoir by a thin diaphragm of metal or melamine according to the required operating pressure. Firing was activated as appropriate by an explosively fired metal dart or by electrically fusing an overlaid matrix of thin copper wires. By initiating a penetration longer than the critical length [96], a diaphragm collapsed virtually instantaneously and the missile itself or in a wooden sabot was accelerated by compressed air along the barrel. Missile velocities up to impact were measured by three independent systems: light beams, fine transverse wires and high-speed cine at 10,000 fps. In the event of target penetration, the subsequent missile velocity was recorded by a similar fine-wire system, high speed cine and two induction loops. As well as visual records of the target's response, analog tape recorders (3dB at 80 kHz) monitored as many as 120 transients from linear displacement transducers and resistance-type strain gauges. Typical diamond sawn cross-sections of damaged reinforced concrete targets are shown in Figure 6.8 where a smooth missile entry occurs because concrete is some 10 times stronger in compression than in tension.

The Horizontal Impact Facility was constructed later to investigate primarily the regulatory compliance of irradiated fuel transport flasks. It had the performance specifications

Maximum projectile energy	2 MJ
Maximum projectile mass	2000 kg
Maximum projectile velocity	250 m/s
Interchangeable barrel IDs	0.5, 1.0, 2.0 m
Post-stressed concrete abutment	1000 tonne

Missile Launcher

A barrel of the Horizontal Impact Facility

Figure 6.7 The Winfrith Missile Launcher and Horizontal Impact Facility [285]

Projectiles were fired from the 0.5 and 1.0 m barrels in the same way as with the Missile Launcher, but firings with the 2 m barrel generally required special arrangements like that shown in Figure 6.9. Here a driver plate guided by four rails first propelled the cradled missile along

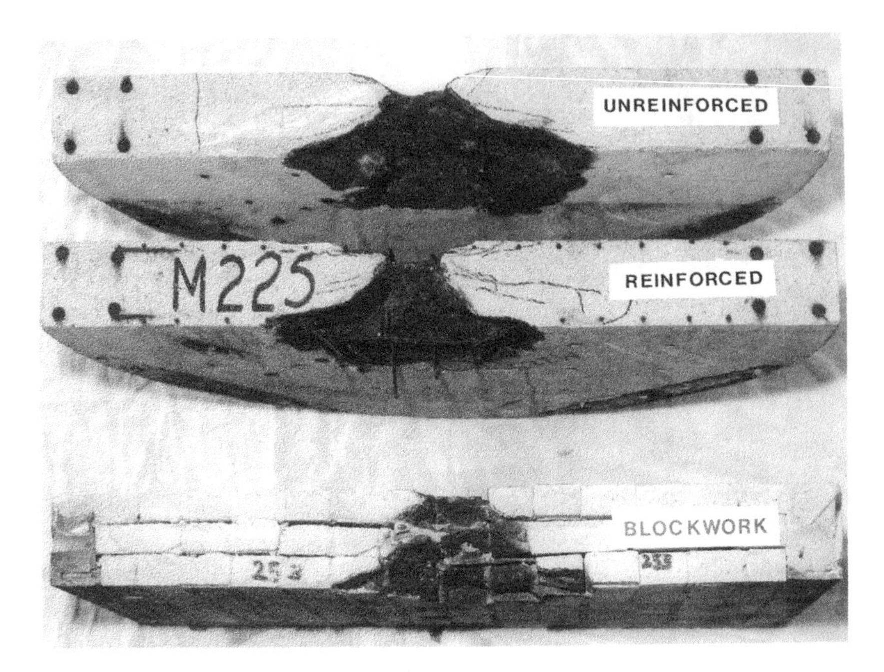

Figure 6.8 Typical Cross Sections of Damaged Concrete Targets

six supporting and aligning rails in the breech. After a short distance the plate and trolley were rapidly arrested by buckling sacrificial lengths of aluminium tubing to release the missile along the barrel. Up to five high-speed cine cameras monitored an impact and an infrared system measured missile velocity. As can be seen by the wires atop of the driver plate in Figure 6.9, transducers were sometimes carried by the missile itself. Any induced pitch, roll or yaw on a projectile evidently reduces the direct energy of an impact, so careful engineering was necessary to restrict these to just $\pm 1°$. Drop height during flight was also constrained to within 120 mm. Quarter-scale replicas of existing steel transport flasks and proposed reinforced concrete designs were successfully tested.

Though valuable in themselves, the principal benefit of these impact experiments lies in underwriting the development of computer simulations. These can now be confidently applied to a wide range of geometries and situations as outlined next in Section 6.4. In passing, the Missile Launcher and Horizontal Impact Facility found commercial applications such as bird strikes on helicopter blades, the survival of

Figure 6.9 Large Replica in HIF Before Firing [285]

air-transport containers in extreme accidents and the effects of various projectiles on glass windscreens.

6.4 COMPUTATIONAL TECHNIQUES AND AN AIRCRAFT IMPACT

Several correlations of experimental data are available for predicting the damage to simple reinforced concrete panels from the impact of hard flat-nosed cylindrical steel billets. Scabbing damage corresponds to dislodgement of a portion of the target's rear face, and perforation has its usual connotation. More than 150 experimental results are available for comparison against these formulae. Neilson [286] establishes that the revised NRDC correlation affords the most accurate predictions for missile penetration depth, and for the scabbing to perforation transition. Though the missile velocity for the onset of

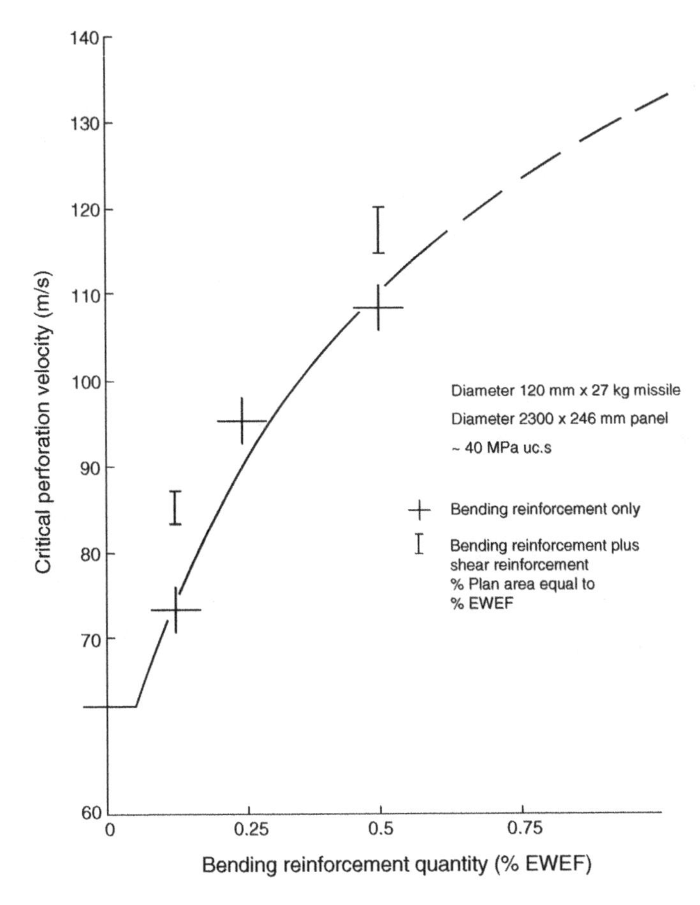

Figure 6.10 Effect of Reinforcement Quantity on the Perforation

scabbing is accurate in 71% of cases, predictions of the perforation velocity are only 57% successful due to the lack of a reinforcement term in the correlation. During an attempted perforation steel reinforcement bars absorb energy by bending and shear, as well as by providing a net to retain broken concrete within the panel. Experimental results from Winfrith [286] in Figure 6.10 confirm a progressive increase in perforation velocity V_p with the amount of reinforcement, and they are correlated by

$$V_p \simeq 132r^{0.27} \tag{6.5}$$

where

r – amount of square mesh reinforcement (% EWEF)

% EWEF – per cent of cross-sectional area occupied by the same square (EW) steel reinforcement just below the surface of each face (EF)

EW – "each way" (horizontal + vertical)

Other Winfrith experiments show that the perforation velocity for steel-faced concrete panels can be predicted by equation (6.5) if the thickness of a rear plate is converted to an equivalent reinforcement percentage. For example, a 1 mm thick rear plate on a 100 mm thick concrete panel corresponds to 1% EW. In the case of a front steel panel, the perforation energy of the composite is the sum of their individual perforation energies. The Ballistics Research Laboratory and Winfrith data are best correlated [286] especially for thicker steel panels by

$$E \simeq 1.44 \times 10^9 (hd)^{1.5} \quad \text{within} \pm 15\% \tag{6.6}$$

where

E — perforation energy (J); h – steel panel thickness (m)

d — missile diameter (m); $hd \lesssim 3.4 \times 10^{-3}$

Ohte et al. [287] confirm that conically-nosed hard missiles perforate targets more readily than flat-nosed ones. For specific missile-target combinations the perforation energy for a hard missile having a 45° half-angle nose is consistently about half that predicted by the BRL formula. With modification of panel thickness and missile diameter as a function of nose-angle in the BRL correlation satisfactory predictions of perforation energy are obtained.

Impacts of irregularly shaped fragments from a disintegrating steam turbine on reinforced concrete, metal panels or major pipe work are important safety issues especially for nuclear plants. Also of real concern is the perforation of a reinforced-concrete containment by aircraft whose geometries and crushing strengths are axially non-uniform. Though correlations are often sufficient for their specific

physical situations having regular simple geometries, applications outside the experimental databases can lead to erroneous predictions. For example a linear extrapolation of the Canfield–Clator correlation [288] for steel projectiles impacting reinforced concrete wrongly suggests that no perforation would occur [289] at velocities less than 150 m/s. Empirical correlations for simple missile and target geometries are therefore inadequate for nuclear plants. Furthermore, the construction of complex replica models would not be cost-effective.
Accordingly the preferred option is the development of comprehensive physics-based computer simulations that are systematically validated by means of replica experiments with simple, but representative portions of the pertinent structures. Finite element techniques [290–293] were developed at Winfrith to pursue this strategy from about 1980.

In essence a finite element calculation divides the region for integrating a partial differential equation into sub-regions (finite elements) that are usually triangular or rectangular, though curved boundaries can be readily accommodated [293]. Values of the required solution at the elements corners constitute the "unknowns," and for small enough elements the solution varies linearly across each element. However, higher-order numerical approximations can be formulated to allow larger mesh sizes [292]. Elliptic equations and the biharmonic equations of structural dynamics are both well suited to finite element techniques because their solutions are equivalent to the minimization of a quadratic functional (e.g., energy). In this form the required values become the unique solution of a linear equation having a positive definite matrix [115]. Galerkin's method is equivalent, if applicable, to a variational approach but the resulting matrix is less sparse and has weaker numerical conditioning [115,292]. Early implementations of the finite element method involved a manual construction and labelling of the mesh that is evidently tedious, error prone and therefore a sizeable portion of the overall cost. Specific computer codes now automate data preparation and the interactive creation of a suitably graded mesh. However, considerable skill and experience are still required to achieve a successful outcome. Numerical solutions with finite elements are presently available for structures involving crushable materials, strain-rate dependent elasto-plastic materials and reinforced concrete with interfacial friction [296]. Dedicated software also assists the interpretation and presentation of computed solutions. The following example illustrates the overall methodology for an aircraft crashing into a reinforced reactor containment.

Prior to unification the large number of military overflights led the Federal German Government to legislate that a nuclear reactor containment must withstand the impact of a Phantom RF-4E aircraft at 215 m/s. However, no such prescriptive requirements apply in the United Kingdom where each site must be separately assessed. Two principal safety concerns are

i. whether a crashing aircraft can perforate the secondary containment or cause unacceptable damage
ii. the nature of vibrations transmitted to the rest of the structure.

Because the crushing strength of an impacting aircraft is so much less than the collapse loading of a reinforced concrete containment, it is therefore a soft missile and deformation of the concrete structure can be neglected in calculating the imposed transient loading. Riera [67] first quantified the situation as part of a safety assessment for the Three Mile Island installation near Harrisburg Airport. His principal assumptions are that an impact produces a region of compacted stationary debris and that the undistorted portion of an airframe continues on towards the target. By Newton's second and third laws of motion the total reaction force $R(t)$ on the building is

$$R(t) = \frac{d}{dt}(mV) = m\dot{V} + \dot{m}V \tag{6.7}$$

where m and V denote respectively the instantaneous mass and velocity of the undistorted portion of the aircraft. If a structure is slowly crushed in a hydraulic press, the involved mass remains constant throughout so the term $m\dot{V}$ is termed the crushing force F_c of the intact portion. With this nomenclature, equation (6.7) becomes

$$R(t) = F_c(t) + \mu V^2 \tag{6.8}$$

where $F_c(t)$now contains an allowance for strain-rate enhancement and μ denotes the instantaneous mass per unit length of the residual airframe. By further assuming that impact velocities remain constant at the approach velocity, the aircraft manufacturer's drawings then

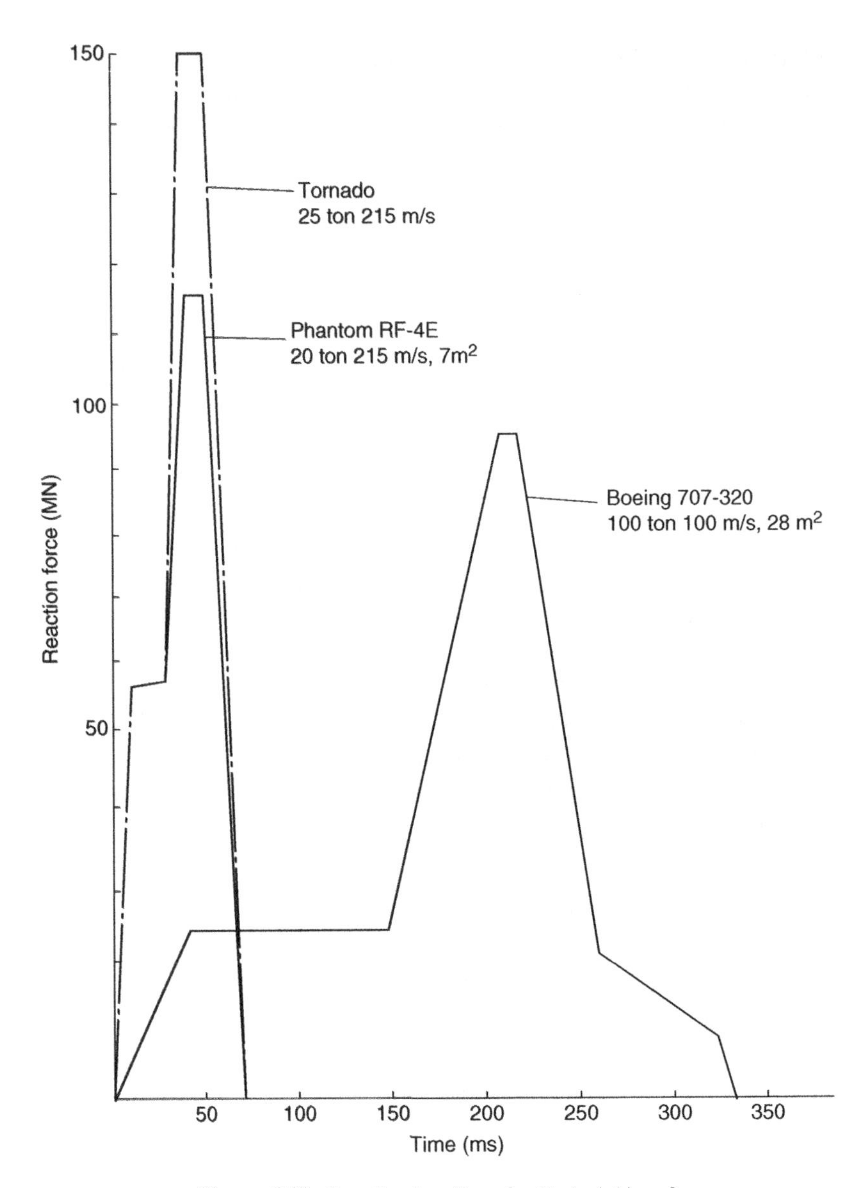

Figure 6.11 Reaction Loadings for Typical Aircraft

provide data for a conservative evaluation of the transient reaction loading on a building. A refinement of the above analysis represents each structurally different longitudinal section of an aircraft, and the loss of mass by dispersed fragmentation should its ultimate compressive

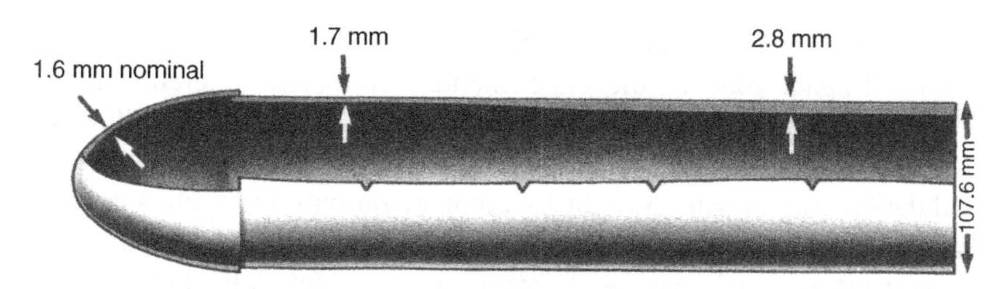

Figure 6.12 Replica Aircraft Model

strength be exceeded [294,295]. These reaction loadings [289] for the impacts of a Boeing 707, Phantom RF-4E, and a Tornado on a rigid structure are graphed in Figure 6.11.

Replica 1/25th scale experiments were performed at Winfrith for an aircraft impacting a proposed reactor containment at around 220 m/s. The aircraft model is shown in Figure 6.12, and the curved reinforced containment was simulated by a flat panel with a massive circumferential ring-beam to provide an edge restraint representative of the remaining structure. Axial mass and stiffness distribution of the model aircraft were engineered to reasonably match the design reaction loading like that in Figure 6.11 The scaled reaction loading in Figure 6.13 formed

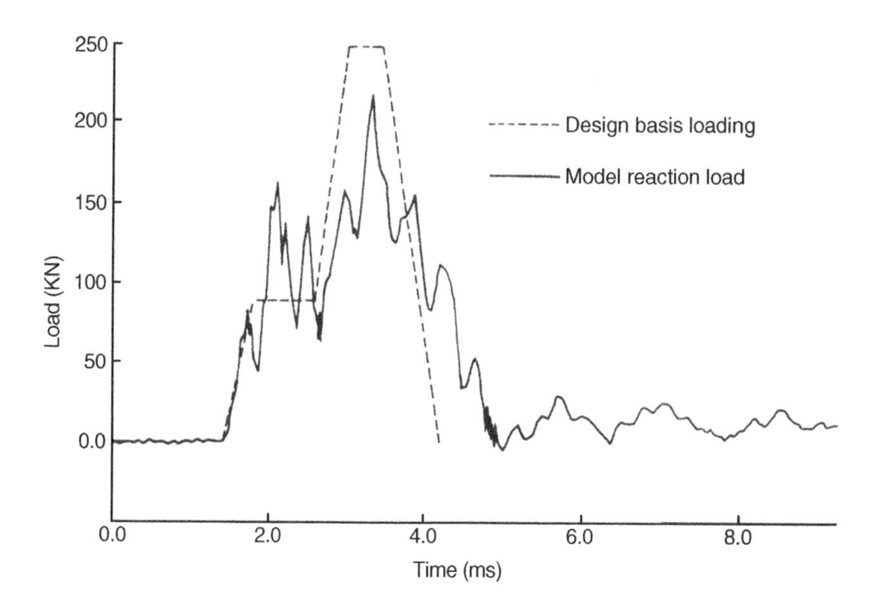

Figure 6.13 Replica Reaction Loading from a Crashing Aircraft

the input to a DYNA-3D finite element code as a transient pressure on the impact zone of the experimental reinforced concrete barrier. A computer simulation then predicted its crack formation etc allowing for interaction between the concrete and its reinforcement. Code validation was as usual sought by comparison with the replica experiments. Though cracking was generally over-predicted [289], good agreement was achieved with regard to transient deflections [296]. Moreover, though model experiments indicated the perforation of a 1 m thick prototype barrier and severe damage at 1.5 m thickness, a 2 m thick containment appeared to be essentially undamaged.

To conclude, the range of Winfrith replica experiments have extensively validated the DYNA-3D code modules for crushable materials, strain-rate-dependent elasto-plastic materials and reinforced concrete with interfacial friction. Nuclear safety assessments can now be confidently implemented with regard to the impact of disintegrating plant items or crashing aircraft on concrete buildings, steel panels or pipe work.

CHAPTER 7

Natural Circulation, Passive Safety Systems, and Debris-Bed Cooling

7.1 NATURAL CONVECTION IN NUCLEAR PLANTS

Following the immersion of a heated surface in a fluid, molecular heat conduction raises local temperatures and thereby reduces fluid densities [208,209]. Buoyancy forces then lift these localized lighter fluid packets to induce a continuous laminar or turbulent flow called natural circulation or natural convection. Industrial research on natural convection began in the 1930s but largely stalled in the mid 1950s because the enhanced heat transfer rates with forced convection (i.e., pumps) enable more compact and therefore more cost-effective plant designs.[1] After the Three Mile Island incident in 1979, natural circulation research restarted to develop passive safety systems [108] that would function even after a total loss of emergency power. An IAEA Conference [298] in 1991 noted that passive safety systems based on natural circulation are a desirable method of enhancing the

[1] Even domestic central-heating boilers now have forced circulation on both flue and water sides.

Nuclear Electric Power: Safety, Operation, and Control Aspects, First Edition.
J. Brian Knowles.
© 2014 John Wiley & Sons, Inc. Published 2014 by John Wiley & Sons, Inc.

simplification and reliability of essential safety functions. Though some new innovative designs for small integrated reactors ($\lesssim$100 MWe) now propose natural circulation as a means of core cooling during normal operation, these have yet to gain formal compliance with European Utility Requirements [59,109].

The principal advantage of natural circulation systems lies in the elimination of active power supplies and pumps, so simplifying plant architecture, maintenance and operation. It also eliminates certain accident scenarios in nuclear power plants, and an increase in channel power intrinsically increases its mass flow rate. On the other hand with a stable forced convective channel an active control system is required to increase its mass flow rate with input power. Specifically, momentum conservation in a vertical channel gives the steady-state pressure gradient [117]

$$-\frac{dP}{dz} = \rho g + \frac{4f}{D}(G|G|/2\rho) + \frac{d}{dz}(G^2/\rho) \tag{7.1}$$

where

ρ – density; g – gravitational acceleration; G – mass flux (kg/m^{2}s)
D – hydraulic diameter; f – Fanning friction factor

Local density and dynamic viscosity μ for a two-phase flow may be calculated as a function of steam quality x from the saturated values as

$$1/\rho = x/\rho_G + (1-x)/\rho_L \text{ and } 1/\mu = x/\mu_G + (1-x)/\rho_L \tag{7.2}$$

and for clean[2] reactor or boiler channels with single or two-phase flow

$$1/\sqrt{f} = -4\log_{10}\left(1.25/Re\sqrt{f}\right) \tag{7.3}$$

[2] Reactor and boiler channels are routinely cleaned chemically during regular maintenance to minimize pumping power so as to promote efficiency.

where

$$Re \triangleq GD/\mu \text{ - Reynolds Number}$$

The driving pressure from modern pumps is only weakly dependent on flow rate, so for present purposes it can be assumed constant. An increase in channel power clearly decreases local density which according to equation (7.1) then increases local pressure gradients. Thus to preserve a constant overall driving pressure, a channel's mass flux must suffer an inappropriate decrease. For this reason boiler feed water flow in conventional forced convective stations is controlled by cascaded throttle valves whose opening increases with demanded output[3] power.

The principal disadvantage of heat removal by natural circulation is that the driving forces from density disparities and gravity are relatively small. Their increase necessitates increases in loop heights and decreases in loop resistances to achieve prescribed heat transfer rates. Because 60 to 70% of the capital costs for nuclear stations reside in civil engineering, the economic viability of plants of order 1 GWe appears questionable. Also larger reactor cores with their smaller bucklings [58] are more susceptible to a spatially unstable neutron flux, and the thermal-hydraulic problems associated with boiling channels are also exacerbated under natural circulation. Specifically, flow stability at economic output powers (exit steam quality) requires the insertion of inlet ferrules to increase the liquid-phase pressure drop as described in Section 3.2. However, this artifice patently reduces the circulating mass flow rate unless loop heights are increased with the penalty of higher construction costs. Moreover, lower mass fluxes and higher steam qualities in two-phase flows encourage a radical reduction in heat transfer that occurs when the liquid phase can no longer "wet" fuel pin surfaces. Under these burn-out[4] or critical heat flux conditions[5] the pressurized fission product gases can locally puncture the fuel

[3] Output demand rather than reactor power to provide anticipatory control; see Section 3.1.

[4] A UK terminology when observed in the SGHWR around 1967 [301].

[5] See Refs. [63,64,117,297]. Burn-out can also occur as a result of nodules of higher enrichment appearing in a later manufactured batch of lower enrichment [301] fuel, or if porous magnetite deposits (crud) on fuel elements have their capillaries blocked by copper salts released from steam condensers [95]. After Dr. G.R. Hewitt identified the mechanism at AERE Harwell, it became known as dry-out, which is a more graphic description than CHF or burn-out for boilers.

cladding whose creep strength [96] progressively weakens with increasing temperature. It is concluded here that heat removal by natural circulation is most cost-effective and reliable for single-phase liquid flows in safety systems removing just the decay heat in Grid-sized reactors ($\gtrsim$ 100 MWe). To meet the United Kingdom's present peak demand of circa 60 GW, economic and environmental considerations favor large stations of around 1 GWe due to the limited availability of suitable sites (Section 4.4) and the existing form of transmission network.

A pioneering paper by Lorenz in 1881 analyzed natural convection heat transfer from a uniformly heated vertical plate in terms of molecular conduction to the neighboring fluid whose reduced density results in its buoyant upward laminar flow [219]. His formula for the corresponding heat transfer coefficient has the form

$$Nu = \text{Function of } Gr(Pr)^{\lambda} \tag{7.4}$$

where

Nu – Nusselt Number; Gr – Grasshof Number; Pr – Prandtl Number

λ – a constant

Later in 1902, Bourinesq showed that when viscosity is negligible (as in very many turbulent flows) the above index λ equals 2. Further investigations [219,303,305] during the twentieth century still generally relate to flat plates or the outsides of horizontal tubes. The earlier experiments reveal that natural circulation can become turbulent with an enhanced heat transfer rate, and this is encouraged in reactor fuel pin bundles by diagrids [268,304]. Heat transfer and pressure drop in tube bundles also depend on their pitch–diameter ratio so the relevant correlations are very much design-specific. Because the gas-side heat-transfer resistance in AGR steam generators is a significant part of the overall primary to secondary-side thermal resistance, specific correlations are necessary for sufficiently accurate predictions of steady-state and dynamic performances [117]. Likewise by virtue of the lower mass fluxes in natural circulation, pertinent experimental heat transfer and pressure drop data are necessary to underwrite proposed safety system designs: especially if the stability of a two-phase channel flow is in question. Thermo-hydraulic transitions in natural circulation are a tractable numerical problem and they can be negotiated like those found under forced convection [64,117]. However, stratified cold sections can be potentially created prior to ECCS operation and these would block decay heat removal by natural circulation, as would an accumulation of nitrogen or hydrogen in the upper U-bend of a steam generator (a loop seal) [306]. Thus the operation of heat removal systems by natural convection must be unequivocally confirmed by experimental rigs. Because full-scale whole plant replicas are impractical, one adopted approach is to adjudge representative cross-sectional sizes and to replicate exactly the differential heights between components so as to preserve gravitational driving forces. A superbly engineered rig relying on this principle has been constructed for example at CEN Grenoble, but the APEX and MASLWR rigs in the United States also scale differential heights by $^1/_4$ or $^1/_3$, respectively [109].

Though natural circulation is proposed for the on-load cooling of Generation IV water reactors [109,298,302], it is installed in presently operational plants only for decay heat removal in potential Severe Accidents. In this context, the separate steam generators of existing PWRs have an important safety role [59,65]. If the primary recirculation pumps in Figure 1.3 were to fail and the system were depressurized,[6] then sufficient coolant injected into the secondary-side boiler inlets is seen to encourage natural circulation in the original flow direction of the reactor circuit. Deaerator or emergency tanks are presently available for this purpose. Moreover, the intrinsic water inventory and that of the ECCS ensure that fuel degradation is a progressive process [66,93,213] so time is available to secure civilian fire engines [65], or to reconfigure the plant to use its steam turbine condensers as heat sinks [59].

An alternative scenario to MFCIs during a Severe Accident in PWRs is a progressively increasing mass of corium on its lower core support structures and at the bottom of its pressure vessel, whose creep strengths [96] are diminished by heat transfer. At high pressures the vessel itself fails first,[7] whereas at lower pressures the supports fail some computed 15 min before the vessel [65]. Steam, fission products and hydrogen from

[6] Temperature-induced density changes in water increase with decreasing pressure [208,209].

[7] Some PWRs have their lower head submerged in water to preserve its creep strength [315].

oxidation of the fuel cladding would then be discharged into the reinforced concrete containment building, whose hydrogen content would further increase from a rapidly developing corium–concrete interaction [65]. A major concern is then a potential rupture of the building from a hydrogen explosion with the consequential atmospheric release of radioactive isotopes. Ejected steam, air and gaseous fission products diminish the likelihood of the initial 7 to 16% by volume of hydrogen [307] to detonate: as do igniters and catalytic recombiners which innocuously dissipate locally concentrated pockets. However, the ejected hot material also threatens the normally well-sealed building by over-pressurization. To restrict the excess pressure to well below its ultimate strength of 5 bar, elevated rings of chemically doped[8] sprays are automatically activated at around 2 bar [66,307]. These produce fine droplets with Sauter diameters of 448 to 544 μm [308], and their large surface area to aggregate volume effects a rapid condensation of steam and ambient cooling, whilst evidently increasing the volumetric concentrations of hydrogen. Herein lies a trade-off problem between spray cooling to counter over-pressurization and the increased risk of a significant hydrogen explosion. It evidently demands serious investigation.

Spray cooling within a reactor containment building creates multidimensional natural circulation flows which to a greater or lesser extent are complicated by

i. Interactive heat transfer between all solid, liquid and gaseous components.
ii. Permanent gases which inhibit steam condensation [219,311]. Also they increase the fugacity [3] of the contents which alters thermodynamic state equations from those of the pure substance [210].
iii. Multicomponent two-phase turbulent flows.
iv. Thermal disequilibrium at liquid–vapor interfaces; see Section 5.5.
v. Dissolved fission products decreasing the vapor pressure of water (boiling point increases)—a fugacity effect.
vi. The efficiency of sump screens in removing core debris from water to be recirculated [310].

[8] See Section 5.8 regarding heat transfer with fine droplets.

Due to the daunting physical and mathematical complexity of a containment's spray cooling process, a combined step-by-step experimental and computer simulation investigation is first necessary to identify the most significant phenomena and then their pertinent interactions. The TOSQAN program is a European initiative along these lines. However, unlike the replica scaling described in Chapter 6, Zuber's time-preserving Hierarchical Two Tiered Scaling Method [307,309] is adopted for this far more complex problem. Though the TOSQAN experiments are not necessarily dynamically similar to reactor-scale processes, the above step-by-step creation of validated computer models eventually results in a reliable simulation for safety assessments at reactor-scale [312,313].

7.2 PASSIVE SAFETY SYSTEMS FOR WATER REACTORS

Passive safety systems consist entirely of passive components, or use active elements in a very limited way to initiate subsequently all passive operations [108]. Manual intervention is excluded in all cases. Because they eliminate multiple pumps with their independent and redundant power supplies some proposed passive systems appear potentially more cost-effective and reliable than current active systems [109]. Accordingly, the IAEA initiated in 2004 an internationally coordinated research project to investigate the performance and reliability of passive safety systems deploying natural circulation for the removal of decay heat after a successful reactor scram (a "hot shutdown"). To provide direction for this experimental and analytical program, the following four degrees of passivity were formulated [108]. The specification for the most stringent Category A passive system is as follows

1. No "intelligent" signal inputs (i.e., activation and operation within the system itself).
2. Neither external power supplies nor forces required.
3. No moving mechanical parts.
4. No moving working fluids.

which are exemplified by

i. Hardened fuel cladding [300]
ii. Containments resistant to excess internal pressures, impacts or seismic activity
iii. Heat removal by thermal radiation or conduction to external structures.

Criteria 1, 2 and 3 are satisfied by Category B systems though moving working fluids are now allowed. Examples include reactor shutdown by the destabilization of hydrostatic equilibrium between its pressure vessel and an external pool of borated water, or containment building cooling by intrinsic natural convection to its internal walls. Category C systems conform to restrictions 1 and 2 but allow moving mechanical parts and fluids. An example in presently operational plants is the pre-pressurized accumulators of borated water which are connected into a reactor circuit by check valves. Under normal operation these are held closed by differential pressures, but in a LOCA the pressure differential reverses and the valves open automatically to shut down the reactor. Category D, the least degree of passivity, allows just battery or gravity-powered active components or intelligent signals to initiate their operation but in no way to control it. Examples include electro-magnetically latched scram rods, and those items listed below for decay heat removal or reactor shutdown:

i. Core make-up tanks of borated water which form part of a natural circulation loop with the reactor.
ii. Borated water injection by gravity at low reactor circuit pressures.
iii. PWR circuit cooling by natural circulation using its steam generators whose secondary circuits become large external cold water sources.
iv. A separate natural circulation heat exchanger whose secondary is a separate large external water tank.[9]
v. Passively cooled core-isolation condensers (BWRs only).

[9] Decay heat removal in PFR was achieved by a NaK air-cooled, natural convective heat exchanger attached to each IHX circuit [314].

vi. A natural circulation loop using water collected in the containment building's sump and the hot reactor core. Any steam produced is vented into and then condensed on its walls, etc.

vii. The spray cooling system for a containment building which is described in Section 7.1.

These Category D passive safety systems and their incorporation into some Generation IV plant proposals are clearly described in Reference 108. However, they are not regarded as preferable to the active systems in presently operational plants. Also the IAEA concludes that the level of understanding and connected code capabilities of the thermal hydraulic phenomena in passive safety systems appear presently to be limited.

7.3 CORE DEBRIS-BED COOLING IN WATER REACTORS

During Severe Accidents quantities of corium as a melt or slurry might slump into the lower head of the pressure vessel.[10] Its sensible and decay heat would then be transferred to the structure whose creep strength progressively decreases with temperature. To mitigate this potential cause of a catastrophic rupture, some PWRs have their lower structures submerged in water [315]. Temperature increases in the pressure vessel are clearly dependent on a debris-bed's morphology, as voidage for example decreases molecular conduction and thereby extends the margin-to or time-to failure. The Three Mile Island incident in 1979 has created the (presently) one and only authentic water reactor debris-bed whose importance to safety assessments motivated an OECD-funded investigation into its relevant characteristics [316].

For a margin-to failure calculation data are required on the physical, metallurgical and radiochemical composition of the debris. Its density, porosity and particle-size statistics broadly characterize morphology, while its metallurgy indicates initial temperatures, melting points, cooling rates and oxidation levels. Radiochemical data completes the required information with regard to decay heat production. "Loose"

[10] Amounts of corium were also found in two steam generators and the pressurizer of TMI-2 [69].

debris with sizes exceeding 150 mm were found to be broadly surrounded by denser material with sizes less than 75 mm. Measured porosities of samples varied markedly between 5.7 to 32% with an average[11] of 18 ± 11%. Chemical analyses reveal that 97% of the debris had the broad composition (U; 70%), (Zr; 13.75%) and (O; 13%) by weight of mixed uranium–zirconium oxides, which supports simulations having a chemically homogeneous melt pool [93,319]. Some samples of these oxide mixtures contained stratified layers of "pores" which are suggested to result from steam or metallic vapors locked in situ as the corium became more viscous prior to solidification. Debris metallurgy indicates gradual cooling rather than rapid quenching, and that single component regions solidified first. The lowest temperature for uranium to dissolve in zirconium is predicted to be around 1760 °C, which is about 1000 °C below the melting point of urania. Metallographic and scanning electron microscope examinations reveal that the maximum temperature for a well–mixed solid solution of uranium–zirconium oxides is between 2600 to 2850 °C. Accordingly, the simulated initial temperature of corium entering the lower head is taken as 2600 °C. Finally, chemical analyses of the debris samples enable the decay heat to be calculated as 130 W/kg and 96 W/kg after 224 and 600 min respectively.[12]

If no MFCIs occur during a Severe Accident, an increasing mass of molten corium would first vaporize any residual water as it slumps into the lower head. Sensible and decay heat would then be transferred by turbulent natural convection into the pressure vessel and by thermal radiation into the degraded structure above [318]. A likely scenario is a multi component liquid pool encased in a growing solidified crust which forms on its free surface and on the vessel's wall. Sohal et al. [315] provide a thorough assessment of available experimental correlations for turbulent natural convective heat transfer between molten corium and solid surfaces. With internal heat generation the recommended correlation takes the form

$$Nu = a(Ra')^b; \quad Ra' \triangleq g\beta q_v L^5 / \alpha \upsilon \kappa \tag{7.5}$$

[11] Higher porosities might have resulted from "Hanging fuel assemblies and control rods strewn about like pick-up sticks on top of a bed of rubble;" see Ref. [69] for the video image.

[12] Noble gas, iodine, and cesium isotopes vaporize out and so are excluded in this calculation.

Table 7.1
Recommended Parameters for Equation (7.5)

Flow Direction	(a; b)	Internal Rayleigh Number Range
Upward	(0.345; 0.233)	[1E10; 3.7E13]
	(0.9; 0.20)	[1E14; 1E17]
Downward	(0.048; 0.27)	[1E12; 3E13]; [1E14; 2E16]
	(0.345; 0.233)	[3E13; 7E14]
Horizontal	(0.6; 0.19)	[1E7; 1E10]
	(0.85; 0.19)	[5E12; 1E14]

where Ra' is the Internal Rayleigh Number and

g – gravitational acceleration; β – thermal expansion coefficient
q_v – volumetric heat generation rate; L – a "characteristic" length
α – molecular thermal diffusivity; υ – kinematic viscosity
κ – molecular thermal conductivity; *a, b* – parameters to match the data

Unlike turbulent flow in pipes and laminar flow aerodynamics, the characteristic length L here is open to more subjective interpretations such as

i. The radius of a pool's top surface
ii. The maximum depth of a pool
iii. The average of 1 and 2 above
iv. The radius of a hemispherical pool equivalent to the volume of the cylindrical one

If the exponent b in equation (7.5) equals 0.2, the characteristic length exactly cancels out. Because the experimental data is best correlated with $b \simeq 0.2$ predicted heat transfer coefficients are quite insensitive to the actual choice. Table 7.1 depicts the recommended values of (a; b) with option 3 to give a regression within about[13] $\pm$ 10% of the data from six water or Freon experiments. Though these experiments

[13] This author's own estimate from graphs in Ref. [315].

Table 7.2
Recommended Parameters for Equation (7.6)

Flow Direction	$(a; b)$	External Rayleigh Number Range
Upward	(0.0923; 0.302)	[2E4; 2E7]
Downward	(0.3; 019) + (0.0462; 0.302)	

involve radically different fluids compared to corium, it should be noted that the burn-out margin for the SGHWR derived from a low-pressure replica-scaled Freon rig accurately matched later measurements derived from a prototype water-cooled rig at 62 bar [63,297]. For situations with Internal Rayleigh Numbers outside those in Table 7.1, the nearest correlation should be used.

An alternative situation to a developing homogenous melt pool is one with a purely metallic layer floating on its upper surface. No internal heat generation occurs within this superficial layer for which turbulent natural convective heat transfer is correlated by

$$Nu = a(Ra)^b; \quad Ra \triangleq g\beta(\Delta T)L^3/\alpha\upsilon \tag{7.6}$$

where *Ra* is the External Rayleigh Number and

ΔT – local temperature difference between the bulk liquid and its boundary

Table 7.2 depicts the recommended values for $(a; b)$ with option 3 as the characteristic length. No well-tested correlation for horizontal flow is apparently available, but as the contact area with a pressure vessel is relatively small, an average of the upward and downward flow coefficients is suggested [315] as adequate.[14]

Molecular heat diffusion into the particulates of a debris bed or in the reactor pressure vessel can be characterized by the axisymmetric form of

$$\frac{\partial T}{\partial t} = \alpha\left[\nabla^2 T + \sigma/\kappa\right] \tag{7.7}$$

[14] Two terms from equation (7.6) added together with these $(a\ ;\ b)$s.

where

α – thermal diffusivity; κ - thermal conductivity
σ – volumetric heat generation rate

Flow through porous media was first quantified experimentally by Henri Darcy in 1856 in connection with aquifers that supplied the civic fountains of Dijon. Under the essentially laminar flow conditions the usually intractable Navier–Stokes equations [256,268] become analytically solvable to generalize Darcy's empirical formula as

$$\boldsymbol{q}_v = -\frac{\zeta}{\mu}\nabla P \tag{7.8}$$

where

$\boldsymbol{q}_v$ – volumetric flow rate (m/s); P – pressure (Pa)
μ – dynamic viscosity (kg/ms); ζ – permeability

The corresponding flow velocity is given by

$$\boldsymbol{V} = \boldsymbol{q}_v/\eta; \quad \eta\text{-porosity} \tag{7.9}$$

with the Reynolds Number

$$Re \triangleq \rho|\boldsymbol{V}|D/\mu \tag{7.10}$$

Here the characteristic dimension D is taken as the smallest sieve-size to allow free-passage for 30% of all particles, and equation (7.8) is valid for $Re \lesssim 10$.

In addition to the above equations and correlations, simulations of debris-bed cooling by the SCDAP/REALP5 code [319] involve

i. The usual two-phase fluid conservation equations
ii. A range of possible debris-bed permeabilities
iii. The oxidation of intact and slumped cladding under re-flooded conditions
iv. The penetration of melted core-plate into existing porous debris, and its effect on heating up the lower head

v. The re-slumping of previously frozen fuel-cladding
vi. The up-take of oxygen and hydrogen under conditions of steam starvation or rapid changes of temperature, etc.

The "stand-alone" development of SCDAP began early in the 1970s to assess the progressive oxidation or melt-down of fuel elements and control rods. Two-phase fluid dynamics patently interacts with these processes, so in 1979 it was merged with the RELAP5 code whose own development had started previously in 1975. Thereafter evaluation and validation of the combination's phenomenological-based modules have been ongoing. It should be recalled that the accuracy of degraded core dynamics is not the usual $\pm$ 10% for engineering design purposes, but it is required only to be demonstrably conservative and bounded. In addition to assessing the margin-to failure of a reactor's pressure vessel, degraded core calculations bound the amount of liquid corium and water in the lower head during a Severe Accident, and thereby the yield of potential MFCIs.

7.4 AN EPILOGUE

It is hoped that readers have concluded that the design and safety of nuclear power plants are soundly established. In this respect, their antagonists sometimes overlook that researchers and operators live in reactor neighborhoods with wives and families.

Spokespersons for non-nuclear organizations frequently assure us that "lessons have been learned:" yet the same misadventures reoccur. This is not the case with the nuclear industry in that the Three Mile Island and Chernobyl [12] incidents are shown herein to have had a tangible impact. Indeed the latter provoked the shutdown of all Russian-designed RMBK reactors in Germany, and also initiated the international requirement for water reactors to have negative reactivity void coefficients. Prior to the earlier US incident, Severe Accidents with fuel melting were regarded as hypothetical or just imaginable. Thereafter internationally coordinated research began into their phenomenological dynamics, and legislation demanded a hierarchical operational structure based on technical qualifications, simulator training and plant experience [59,65,91]. In this respect, note that human error aggravated or precipitated both incidents. It is also contended herein that the Fukushima incident resulted from a site-planning error rather than from a flawed nuclear technology. Because these reactors lay on the unsheltered East Coast, the enormous tidal surges were able to swamp the emergency power supplies for the ECCS. If they had been sited on the West Coast, as some other Japanese plants, they would have been shielded from tsunamis and their successfully initiated neutronic shutdowns by scram rods likewise maintained in spite of the Richter-scale 9 earthquake.

Renewable energy sources, especially windpower, were incipiently greeted with public approbation for their perceived potential to decelerate global climate change. Indeed a UK government ratified an international agreement for particularly low national carbon emissions believing that renewable energy sources would deliver its promise. Now however the media and an often vociferous public raise concern over renewables' green credentials and their ability to provide a secure UK electricity supply. If the proposed 18 GW of wind power are lost by not uncommon nationwide high-pressure weather, then despite the proposed construction of a European Supergrid [38,39,43], it is argued that the necessary back-up would not always be available. Specifically, Northern Europe has the heaviest industrialization with peak demands in the same winter solstice as the United Kingdom, and their existing aged stations lack the requisite margins. Moreover, Germany's intention to shut down all its nuclear plants by 2020 clearly exacerbates competition for any available power by the privatized utilities: even if it were to be available. Finally, wind turbines annually produce only some 20% of their rated outputs [36] which intensifies this competition, worsens their economics and necessitates reliable back-up by gas-fired or nuclear stations whose capacity factors are between 80 and 90%.

Actual or potential environmental impacts have been identified for all commercially viable renewable energy sources. For instance the Isle of Thanet wind farm of a nominal 300 MW has a visual impact of over some 3500 ha,[15] whereas the relatively unobtrusive SGHWR generated 100 MWe with over 60% reliability on a total of just $3\frac{1}{2}$ ha which included car parking areas. The "largest whirlpool in the world" is created when sluices of the La Rance tidal barrage open to capture a rising tide. Rather than being viewed in terms of estuarial damage, it has become a tourist attraction and shortly after construction EDF quietly

[15] 1 ha approximates to 1 football pitch.

switched to nuclear power. To place the heroic loss of emergency workers' lives at Chernobyl in perspective, the catastrophic failure of the Banqiao hydroelectric dam in 1975 resulted in 171,000 fatalities [11], whilst a multi-national IAEA investigation [13] concluded that no subsequent medical conditions could be directly attributed to this crassly initiated and managed nuclear incident. Radioactive releases at Three Mile Island have been calculated to induce just an additional one or two thyroid cancer presentations during the following 20 years, [66] and there were no direct fatalities.

The necessary back-up for renewable generation can be provided only by fossil or nuclear stations. Until carbon capture [56] becomes viable, neither coal nor partially fossilized lignite is likely to be deployed as fuel. Over the past 2 years, there has been a marked shift in the prospects for gas as a result of the enormous quantities of shale gas made available by fracking. Though exploration in the northwest of the United Kingdom has apparently identified commercial quantities, its large-scale exploitation is under review due to concerns over minor earth tremors and possible aquifer pollution. These anxieties appear less relevant in the far less densely populated United States, which has become a potential major exporter of liquefied gas rather than a previous importer. Since 1992 the high thermal efficiencies ($\simeq$50%) and favorable cash flows of combined cycle gas turbine (CCGT) plants have resulted in their increasing UK deployment [51,52]. However, a KPMG report [52] concludes that the supply of UK electricity by a continuing investment in CCGT plants alone could not meet government obligations on carbon emissions. In addition, the country is not self-sufficient in gas and so is vulnerable to external political unrest. On this basis nuclear power is argued herein to be a necessary component in a "mix" to guarantee the United Kingdom a secure and reliable electricity supply within its emissions obligations.

As well as largely predictable daily and seasonal changes, there are rapid unpredictable power variations on a Grid network such as the start-up of a 1 MW main-line locomotive or the disconnection of a 1000 MW station due to lightning, a bird strike or component failures [80]. Transient differences between instantaneous power generation and that consumed are shown to create network-wide common frequency fluctuations about the nominal UK value of 50 Hz, and for the explained technical and statutory reasons these fluctuations must be within $\pm$ 0.5 Hz. Due to operational rate constraints rapid unexpected changes in power demand cannot be met by modulating the primary energy sources, but only from the thermal energy stored in components of the load-following (Coupled) stations and the rotations of all synchronized motors and generators [80,117]. In this respect gas-fired CCGT stations, preferably with steam drums, are more flexible than PWR and BWR designs whose slower power changes are imposed to achieve economic fuel cycles[16] and the intervention of safety systems. Accordingly, nuclear stations operate in the decoupled mode to supply the more slowly varying and largely predictable base load: as consistent with their larger initial capital costs. The UK mix of CCGT, nuclear and wind power generation as outlined by the government on May 20, 2012, hinges at the time of this writing on consumer price guarantees to attract private equity investment.

An atmospheric release as aerosols of radioactive Iodides and Caesium in the size range 1 to 5 μm is identified as the principal hazard in nuclear power production. However, the dispersed mass would be markedly reduced by dissolution in the large quantities of reactor coolant present [104]. For instance just some 16 curies of the 3 to 5 million in the Three Mile Island reactor were released due to absorption in the water or vapor present in the containment building and its sump [66]. Since then enhanced safety systems, statutory operating protocols and pre-planned public evacuation schemes have been put into place [59,65,91]. It is also required that the aggregate probability of all Severe Accidents with or without an atmospheric release must be no greater than 10^{-7} per operating year throughout a plant's lifetime. Reactor safety assessments such as Farmer's [157] specify a particular station to be safe if and only if the statistically expected increase in cancers from all Severe Accidents during its operational life is orders of magnitude less than from natural causes. Radiological dose rates in such assessments involve variations in local population density and wind-direction statistics. However, they over-predict fatalities because

i. Induced cancers from a radiation dose are calculated from linear extrapolations of data from Japanese A-bomb survivors. Consequently there is no allowance for the natural repair mechanisms of the human body that become effective at

[16] By restricting differential thermal expansion between fuel pellets and clad.

much lower doses [66]. For example, the background radiation in the granite city of Aberdeen is three times that in London: yet there is no statistically significant increase in comparable cancers.

ii. Some 1770 cases per million of natural thyroid cancers present annually in the United Kingdom of which 80 to 90% are successfully treated by surgery [163,164].

It is also contended that safety assessments are biased against nuclear power because

a. Risks are assessed without attendant benefits: like some medical vaccinations [169].

b. Humans peculiarly accept much larger self-imposed risks than risks externally imposed, and a cancer risk is to be avoided no matter how slender. Specifically nuclear power is often rejected despite the orders of magnitude greater fatal risks from natural cancers, road traffic accidents and tobacco smoking (see Section 4.3).

Public concern remains over the storage of nuclear waste. Reference [320] details its classification by activity levels and the appropriate storage technologies. Unlike the atmospheric release of dioxin at Bhopal with an indeterminate active life, radioactivity in nuclear waste reduces to that of mined ores after about 7000 years [320]. To obstruct any environmental release during and beyond this period, a well-researched multi-barrier technology has been developed for high level (highly active) waste. After an initial storage in "ponds" to reduce decay heat the contents of intact fuel pins are glassified before being concreted within copper-clad stainless steel drums. Finally these are embedded in Bentonite clay before being stored in deep igneous rock formations that allow their subsequent inspection or retrieval. The natural fission reactor at Oklo [321] about 1800 million years ago demonstrates that impervious igneous rock strata alone can retain fission products for well over the required time span. With regard to public opinion, 80% of the populations in Sweden's Forsmark and Oskarsham towns voted in favor of local storage facilities in their neighboring igneous rock tunnels [32].

Commercial nuclear power generation is frequently wrongly associated with offensive weapons due to the military antecedents of nuclear fission. In fact the economic viability of nuclear power necessitates the fission of just so much U-235 (fuel burn-up) that fissile Pu-239 transmuted from U-238 becomes itself transmuted into the α-emitter Pu-240 in concentrations greater than 7%. Consequently the inseparable plutonium mixture is then unsuitable for weapons. Indeed nuclear power was described in the pioneering days of Calder Hall as "The Peaceful Use of Atomic Energy".

Power generation by thermal reactors involves less complicated (expensive) engineering than by fast reactors, which for example require intermediate heat exchangers to isolate further the primary circuit's sodium from steam generators. There is presently no foreseeable shortage of mined uranium, so the fertile-to-fissile fuel-breeding feature of fast reactors has neither strategic nor immediate economic benefit. Accordingly thermal reactors have become the worldwide choice. Section 1.8 shows that water reactors are intrinsically the most compact designs, and because 60 to 70% of a nuclear station's capital costs reside in civil engineering [74], they are also more cost-effective than grosser AGRs. Though BWRs need neither separate steam generators nor pressurizers, Section 2.3 shows that boiling channels can be unstable, so this apparent advantage is eroded by their lower linear fuel ratings that are necessary to present fuel damage by burn-out [63,64,297]. Other relative disadvantages of BWRs are identified so that PWRs have become the more favoured worldwide choice. Finally, SGHWRs require enrichment of both fuel and moderator (deuterium) with the former necessary to achieve a negative reactivity void coefficient [61]. Consequently, economics have led to their discontinuation despite their reliability and safety having been proved over 1966–90.

References

1. National Grid, plc; "Annual Report and Accounts 2009/2010".
2. Pippo, Di. "Geothermal Power Plants: Principles, Applications, Case Studies, and Environmental Impact," Oxford Butterworth –Heineman, 2007.
3. Zemanski, M. W. "Heat and Thermodynamics," McGraw Hill, 1951.
4. "Hydroelectric Power Water Use," *Water Science for Schools*, US Geological Survey.
5. Wikipedia.org/wiki/hydroelectricity.
6. "Business Supplement," *Sunday Times, (August 29, 2010).*
7. "Electricity Production Statistics 2008," eia.doe.gov.
8. "A Looming Threat to Russia's Mighty Rivers," www.pacific environment.
9. "One dam mistake after another leaves $4.4 bn bill"; www.smh.com.au
10. www.secret-scotland.com/pitlochrysalmon-ladder.html
11. "The Catastrophic Dam Failures in China"; August 1975. www.sju.edu/faculty/watkins/aug1975.au
12. Grimstone, M. C. "The Chernobyl Accident: A Review," *AEA Technology Report* (1991).
13. "International Chernobyl Report: An Overview," *IAEA Wien* (1991).
14. www.actionmaster.com (Sept. 2010).
15. Franzen P. "The Fundamental Principles of Heat Transformers," *Gesellschaft für Systemtechnik* GmbH (1979).
16. Crowther, J. A. "*Ions, Electrons and Ionizing Radiations*", Arnold, 1952.
17. "Photovoltaic Electricity," wikipedia.org/wiki/photoelectricity.
18. bbc.co.uk/weather (Sept 2010).
19. "Solar Production," wikipedia.org/wiki (Sept 2010).
20. www.uklanddirectory.org.gov.uk (Sept 2010).
21. Lerner, E. J. "Low-head Hydro Power," *IEEE Spectrum* (Nov., 1980).
22. "Pros and Cons for Wave and Tidal Power," oceanenergycouncil.com.
23. "Tidal Technologies Overview – Research Report 2" for the UK Sustainable Development Commission (2007).

Nuclear Electric Power: Safety, Operation, and Control Aspects, First Edition.
J. Brian Knowles.
© 2014 John Wiley & Sons, Inc. Published 2014 by John Wiley & Sons, Inc.

24. *"Reeds Nautical Almanac"*; www.reedsnauticalalmanac.co.uk
25. "US Nuclear Availability 2007 and 2008," eia.doe.gov.
26. "The Economics of Nuclear Power," World Nuclear Association (April 2010).
27. "La Rance"; energylibrary.com.
28. Warne, D. F. and Calnan, P. G. "Generation of Electricity from the Wind," *Proc. IEE* ***124****, No 11R*, (1997) 963–985.
29. Koenroads, A. J. T. M. et al., "Some Aspects of the Dutch National Research Programme for Wind Energy," *Electrical Engineering Department Report* from Eindhoven University; The Netherlands (1979).
30. "Isle of Thanet Windfarm," bbc.co.uk/news (Sept. 24, 2010).
31. "Isle of Thanet Windfarm," *Power Technology*, (Sept. 2010).
32. Personal communication from Decommissioning Department – www.lesley cox@research-sites.com.
33. "Grid-tied Inverters," wikipedia.org/wiki/grid tied inverter
34. "Summary Report of DoE Hightech Inverter Workshop," (Jan. 2005).
35. Say, M. G., "*The Performance and Design of Alternating Current Machines*", Pitman (1955).
36. "Capacity Factors for Windpower," wikipedia.org/wiki/wind.
37. "Energy Security – Looking to the Future," *Shares*, Nov. 6 (2008).
38. "Supergrid Paves the Way for Windpower Expansion," airchim.org/factsheets/North.
39. "Supergrid Declaration," claverton-energy.com/north-sea.
40. Hernen, B. "The Dinorwic Pumped Storage Scheme," *IEE Electronics and Power*, (Oct., 1977).
41. Miford E. "UK Prepares to Realise Its Offshore Wind Potential," *Renewable Energy World* (Oct. 2010).
42. "Wind map"; www.metoffice.gov.uk.
43. "Renewable Energy Supergrid Coming to Europe," *Alternative Energy* (Jan. 13, 2010).
44. "Supergrid Gets Serious, but Does It Rely Too Much on Norway?" blogs.ft.com (March 8, 2010).
45. "Utility-scale Efforts," *Shares*, (Sept 2, 2010).
46. Porritt J. "Is Nuclear the Answer?" w.w.w.sd-commission.org.uk.
47. "Sustainable Development Commission Disbanded," bbc.co.uk/news (July, 22 2010).
48. "How Electricity is Generated in the UK," hi-energy.org.uk.
49. "What is a Carbon Footprint?" www.parliament.uk/documents/post
50. "DOE. Fossil Energy: Overview Fluidised Bed Technology," (Feb. 2009), fossil.energy.gov/combustion/fluidised bed.
51. "Brown-Boveri Combined Cycle Power Plants" – Series KA9, KA11 and KA13; Brown-Boveri Publication No. CH-KW 1275 87E.

52. "Securing Investment in Nuclear in the Context of Low-Carbon Generation," *KPMG Report*, (July 2010).
53. "Coal Reserves," wikipedia.org/wiki/coalreserves.
54. Ballentine, C. "How to Capture Carbon," www.sci.manchester.ac.uk
55. "USA Coal Reserves," eia.doe.gov/fuelcoal
56. Interview on "Countryfile," TV Program, BBC1, (Oct 17, 2010).
57. *Digest of UK Energy Statistics* (DUKES).
58. Glasstone, S. and Edlund, M. C. "*Elements of Nuclear Reactor Theory*", Van Nostrand, 1965.
59. "PWR Degraded Core Analysis," Committee Report Chaired by John Gittus, UKAEA ND R-610(S) (1982).
60. Hirsch, Sir Peter, "The Fast Reactor: Perspective and Prospects," *UKAEA Atom* **325**, 242–251 (1983).
61. Gibson, I. H. "Light-water Reactors," Graduate Training Notes AEEW (1976).
62. "Nuclear Power Reactors," UKAEA Information Services (1987).
63. Butterworth, D. and Hewett, G. F. (eds), "*Two-Phase Flow and Heat Transfer*", Oxford University Press, 1977.
64. Knowles, J. B. and Robins, A. J. "Heat Transfer Regimes and Transitions in Boiler Dynamic Models," *BNES Conference on Boiler Dynamics and Control in Nuclear Power Stations* (1973).
65. "German Risk Study Nuclear Plants – a Summary"; *Gesellschaft für Reaktor Sicherheit Report* 74 (1990).
66. Davis, L. M. "The Three-Mile Island Incident," *AERE Report* (1979).
67. Riera, J. D. "On the Stress Analysis of Structures Subjected to Aircraft Impact Forces," *Nuclear Eng. and Design*, **8**(4), 415 (1968).
68. Barr, P. et al., "Replica Scaling Studies of Hard Missile Impacts on Reinforced Concrete" in *Concrete Structures under Impact and Impulsive Loads*, edited by Plauk, B. A. M., Berlin (1982).
69. Fischetti, M. A. "TMI plus 5: Nuclear Power on the Ropes," *IEEE Spectrum* (April 1984).
70. AMEC plc Annual Reports and www.amec.com.
71. "Winfrith SGHWR," research-sites.com.
72. "Nuclear Power in Sweden," *World Nuclear Association Report* (Sept 2010).
73. "Nuclear Power Plants – Worldwide," European Nuclear Society (Oct 2010).
74. "Economics of Nuclear Power," World Nuclear Association, (April 2010).
75. *Sunday Times* (Oct 24, 2010).
76. Seelmann-Eggebert, W. "Nuklidkarte," KfK (1974).
77. Lefschetz, S. (ed.), "*Contributions to the Theory of Non-Linear Oscillations*", Princeton, 1950 to 1960.
78. Aizerman, M. A. "On the Problem Concerning the In-large Stability of Dynamic Systems," *Uspekhi Mat. Nauk* **4**, 187 (1949).

79. Knowles, J. B. "A Comprehensive, yet Computationally simple, Direct Digital Control System Design Technique," *Proc. IEE* **125**, 1383–1395 (1978).

80. Knowles, J. B. "Principles of Nuclear Power Station Control," *BNES Journal* **15**, 3, 225–236 (1976).

81. Hughes, F. M. "Analysis and Design of a Nuclear Boiler Control Scheme," in *Design of Modern Control Systems*, Bell, D. J. et al, (eds) IEE Series **18**, 106-125 (1982).

82. Munro, N. "The Inverse Nyquist Array Design Method" in *Design of Modern Control Systems*, Bell D. J. et al, (eds) IEE Series **18**, 83–105 (1982).

83. Potter, R. "A Review of Two-Phase Flow Instability: Aspects of Boiler Dynamics," *BNES Conference on Boiler Dynamics and Control*, London (1973).

84. Butterfield, M. H. and Paravacini de, T. P.– unpublished work, C and I Division at AEEW (1968).

85. Hicks, E. P. and Menzies, D. C. "Theoretical Studies on the Maximum Fast Reactor Accident," *ANL-7120* (1965).

86. Berthond, G., Jacobs, H., and Knowles, J. B. "Large Scale Fuel–Sodium Interactions Performed in Europe," *O-ARAI Conference*, Japan (June 1994).

87. Smith, B. L., Washby, V. and Yerkess, A. "SEURBNUK-EURDYN: First Release Version," SMiRT 9, Lausanne (1987), Paper E6/6.

88. Hoskins, N. E. and Lancefield, M. J. "The COVA Programme for Validation of Computer Codes for Fast Reactor Containment Studies," *Nuclear Eng. and Design* **46**, 17 (1978).

89. Bird, M. J. "An Experimental Study of Scaling in Core-Melt/Water Interactions," *National Heat Transfer Conference*, Niagara Falls (August 1984).

90. Buxton, L. D. and Benedick, W. B. "Steam Explosion Efficiency Studies," NUREG/CR-0947-SAND79-1399 (1979).

91. "Requirements for New Operating Licences,"NUREG-0694 (1980).

92. "EDF Post-TMI Programme," *Framatome* (1979).

93. Siefken, L. J. et al. "SCDAP/RELAP5 Modeling of Heat Transfer and Flow Losses in Lower Head Porous Debris," *Idaho National Engineering and Environmental Laboratory Report* (1998).

94. Yong-Hoon, K. and Kune, Y Sah "Sensitivity Analysis for Maximum Heat Removal from Debris in the Lower Head," *J. Korean Nuclear Soc.* **32**, Vol 4, 395–409 (2000).

95. Macbeth, R. V. "Boiling on Surfaces Overlayed with a Porous Deposit: Heat Transfer Rates Obtainable by Capillary Action," *AEEW Report* 711 (1971).

96. Cottrell, A. H. "*The Mechanical Properties of Matter*", Wiley, 1964.

97. "A Study of the Risk Due to Accidents in Nuclear Power Plants," German Federal Ministry for Research and Technology, Bonn (1979).

98. "Steam Explosions in Light Water Reactors," *Swedish Government Report DsI* (1981).

99. Cain, S. and Haber, L. "Ensuring Nuclear Plant Safety Following a Loss of Coolant Accident," *Power-Gen* (Sept 2010).
100. Rao, D. V. et al., "Knowledge Base for the Effect of Debris on PWR Emergency Core Cooling Sump Performance"; NUREG/CR-6808; LA-UR-03-0880 (2003).
101. Williams, D. C. et al. "CONTAIN Code Analysis of Direct Containment Heating (DCH) Experiments," Sandia Report 095-1035C (1995).
102. Kendall, K. C. et al. "Experimental Validation of the Containment Codes ASTARTE and SEURBNUK," *5th Int. Conference SMiRT*, Berlin (August 1979).
103. Kendall, K. C. et al. "A SEURBNUK-EURDYN Calculation of a COVA Experiment Representative of a Prototype Fast Reactor Design," *6th Int. Conference, SMiRT Paris* (1981).
104. See *Nuclear Technology 53* No 2 (May 1981) for many papers on the physical-chemical processes that would restrict the dispersions of fission products after a Severe Accident.
105. Barr, P. unpublished work at AEEW (1981).
106. Barr, P. et al. "An Experimental Investigation of Scaling of Reinforced Concrete Structures under Impact Loading," *Institute of Structural Engineers/BRE Seminar on Dynamic Modelling of Structures*, Watford (1981).
107. Neilson, A. J. Private communication (1993).
108. Cleveland, J. and Choi, J. H. "Passive Safety Systems and Natural Circulation in Water Cooled Nuclear Plant," IAEA-TECDOC-1624 (2009).
109. Reyes, J. and Cleveland, J. "Natural Circulation in Water Cooled Nuclear Power Plants," IAEA-TECDOC (2005).
110. Halmos, P. R. "*Finite Dimensional Vector Spaces*", Van Nostrand, 1963.
111. Taylor, A. E. "*Introduction to Functional Analysis*", Wiley, 1964.
112. Dieudonné, J. "*Foundations of Modern Analysis*", Academic Press, 1963.
113. Phillips, E. G. "*Functions of a Complex Variable*", Methuen, 1954.
114. Titchmarsh, E. C. "*The Theory of Functions*", Oxford University Press, 1939.
115. Bodewig, E. "*Matrix Calculus*", North Holland, 1959.
116. Stevens, B. L., private communication (Jan. 2012).
117. Knowles, J. B. "*Simulation and Control of Electrical Power Stations*", Research Studies Press/Wiley, 1990.
118. Knowles, J. B. "Transient Heat Release from Steam Drum Metal," *CEGB Symposium on Conventional Boilers*, London (1980).
119. Jaeger, J. C. "*An Introduction to the Laplace Transformation*", Methuen, 1955.
120. McLachlan N. W. "*Complex Variable Theory and Transform Calculus*", Cambridge University Press, 1955.
121. Gantmacher F. R. "*The Theory of Matrices*", Vol. 1 and 2, Chelsea, 1959.
122. Owens D. H., "*Feedback and Multivariable Systems*", Peregrinus, 1978.
123. Rosenbrock, H. H. "Design of Multivariable Control Systems Using the Inverse Nyquist Array," *UMIST Control Centre Report* No 48 (1969).

124. Knowles J. B. "*Direct Digital Control Systems*", Research Studies Press/Wiley, 1994.

125. Knopp, K. "*Theory and Applications of Infinite Series*", Dover, 1990.

126. Massera J. L. "Contributions to Stability Theory," *Annals of Mathematics* **64**, 182–205 (1956).

127. Smith, I. C. and Wall, N. "Programmable Electronic Systems for Reactor Safety," *UKAEA Atom* 10–14 (1989).

128. Knowles, J. B. "A Brief Guide to Deterministic Frequency Response Measurements," *AEEW Memo* 1547 (1977).

129. Guilleman, E. A. "*The Mathematics of Circuit Analysis*", Wiley, 1951.

130. Taylor P. L. "*Servomechanisms: An Introduction to the Theory and Practice of Closed Loop Control Systems with an Account of Data Transmission And Computation*", Longmans, 1960.

131. James, H. M. et al. "*Theory of Servomechanisms*", MIT Radiation Laboratory Series; **25** McGraw (1947).

132. Butterfield, M. H. and Paravacini de, T. P., unpublished work at AEEW (1975).

133. Hughes, F. M. "Analysis and Design of a Nuclear Boiler Control Scheme," in *Design of Modern Control Systems*, IEE Series **18**, 106–125 (1982).

134. Rosenbrock, H. H., "*Computer-Aided Control System Design*", Academic Press (1974).

135. Van Der Weiden, A. J. J. "Inversion of Rational Matrices," *Int. J. Control* **25**, No 3, 393–402 (1977).

136. Emre, E. et al. "On the Inversion of Rational Matrices," *Trans IEEE Circuits and Systems* **21**, 8–11 (1974).

137. Munro, M. and Bowland, B. J. "The UMIST Computer-Aided Design Suite: Users Guide," UMIST Control Systems Centre Report (1980).

138. Munro, M. and Engell, S. "Regulator Design for the F100 Turbofan Engine," *IEEE Conf. on Control and its Applications*, Warwick University p. 380–387 (1981).

139. Winterbone, D. E. et al. "Design of a Multivariable Controller for an Automotive Gas Turbine," *ASME Gas Turbine Conf.* Washington, Paper 73-GT-14 (1973).

140. Kidd, P. T. et al. "Multivariable Control of a Ship Propulsion System," *Proc. 6th Control Systems Symposium*, Canada (1981).

141. Knowles, J. B. and Farrier, D. R. "A Grid Disconnection Strategy for a Benson Steam Cycle in a Fast Reactor Power Plant," *Boiler Dynamics and Control BNES Conference*, Bournemouth (1979).

142. Knowles, J. B. and Collins, G. B. "Fast Reactor Dynamic Performance Optimisation," *BNES Conference on Sodium Cooled Reactors*, London (1977), Paper 21.

143. Hewitt, G. F. and Shires, G. L. "*Process Heat Transfer*", CRC Press, 1979.

144. Strickland, E. "Twenty-Four Hours at Fukushima," *IEE Spectrum* (Nov. 2011).

145. Moxon, D. "A Dynamics Programme for Nuclear-Thermal-Hydrodynamic Behaviour of Water Cooled Reactors," *AEEW Report* 441 (1966).
146. Knowles, J. B. "Molten Fuel–Coolant Interactions," *European Commission WAC Group Meeting*, Obninsk (1993).
147. Peek, F. W. "The Law of Corona and the Dielectric Strength of Air," *Trans AIEE*, **30**, 1899 (1911).
148. Bevan T. "*The Theory of Machines*", Longman, 1953.
149. Golding, E. W. *Electrical Measurements and Measuring Instruments*, Pitman (1955).
150. "Real Time Operational Data," www.National Grid Electricity.
151. Cummins, J. D. and Green, M. A. "Statistical Characteristics of Observed Grid Frequency Variations using the Programme SAMPAN," *AEEW Report*, 835 (1972).
152. Horst, D. M. "A Discussion of the GE/Bechtel Prototype Large Breeder Reactor," *BNES Conference on the Optimisation of Sodium-Cooled Fast Reactors*, London (1971), Paper 30.
153. Ostrovski, Yu. I. "An Extremum Controller for the Turbine Drilling of Oil Wells," *Auto-matika i Telemakanika*, **18**, 852 (1957).
154. Knowles, J. B. "A Contribution to Computer Control," Ph.D. Thesis, UMIST (1962).
155. Jowett, J. Private communication (2007).
156. Lord Rothschild,"Risk," BBC (Richard) Dimbleby Lecture (1978).
157. Farmer F. R. "Siting Criteria: A New Approach," IAEA Paper SM-89/34 Vienna (1967) with an Appendix by J. R. Beattie.
158. Hörtner, H. "German Risk Study: Results of Event Tree and Fault Tree Analysis," *Proc. Int. Conf. on Probabilistic Safety Assessments and Risk Management, Zürich* **2**, 419, TUEV (1987).
159. Jaynes, E. T. "*Probability Theory: The Logic of Science*", Cambridge University Press, 2003.
160. Stigler, S. M. "*The History of Statistics*", Harvard University Press, 1986.
161. Int. Committee on Radiological Protection , "The Evaluation of the Risks from Radiation," **1**, Publication 8, Pergamon (1966).
162. Inst. Committee on Radiological Protection , "Limits of Intakes of Radionuclides by Workers," Publication 30 Part 1, Pergamon (1978).
163. Saunders, P. "The Effects of Radiation on Man," *UKAEA Atom* **298**, 198 (1981).
164. "UK Thyroid Cancer Deaths 2008," www.cancerresearch.org/print/%2520
165. "UK Road Accidents 2009," www.dft.gov.uk/pgr/statistics
166. "UK Smoking Deaths," www.ic.nhs.uk/pubs/smoking
167. Rasmussen, N. C. et al. "Reactor Safety Study. An Assessment of Risk in US Commercial Reactor Plants," WASH-1400, USNC Report NUREG-(19)75/014 (Oct 1975).

168. Dunster, H. J. "An Iconoclastic View; The Assessment of the Risks of Energy," *UKAEA Atom* **303**, 2 (1982).
169. Salk, J. An interview with B. Levin on BBC2 (June 1982).
170. Bitt, H. "Prevention of Uncontrolled Dissemination of Radioactive Aerosols," IAEA-CN-70/73.
171. The significant research at UKAEA Harwell is reported in the *Journal of Aerosol Science* by Dr. C. Clements, et al.
172. Pasquill, F. "The Estimation of the Dispersion of Windborne Material," *The Meterological Magazine* **90**, No 1063, 33 (1961).
173. Cramer, H. *Mathematical Methods of Statistics*, Princeton, 1954.
174. ICRP Committee No. 1, "The Evaluation of Risk from Radiation," Publication No. 8, Pergamon (1966).
175. Beattie, J. R. and Bryant, P. M. "STRAP: A Site Risk Assessment Programme," unpublished work at UKAEA Risley.
176. Winkler, R. L. "*An Introduction to Bayesian Inference and Decision*", Holt-Rinehard and Winston, Inc. (1972).
177. Broadly, D. Private communication NNC (1982).
178. Rippon, S. "A PWR for the UK," *UKAEA Atom* **307**, 100 (1982).
179. "A Technical Outline of Sizewell B, The British Pressurised Water Reactor," CEGB (1982).
180. Akson, N. "Selected Examples of Natural Circulation for Small Break LOCA and some Severe Accidents," *IAEA-ICTP Conference on Natural Circulation in Water-Cooled Nuclear Power Plants*, Trieste (2007).
181. Alsmeyer, H. "BETA-Experiments in Verification of the WECHSL-Code: Experimental Results on Melt-Core Interaction," *Nuclear Engineering and Design* **103**, 115 (1987).
182. Alsmeyer, H. "Modellentwicklung zur analytischen Beschreibung von Kernshmelzunfällen," PNS-Jahresbericht KFK4100 (1987).
183. Schöck, W. et al. "The Demona Project, Objectives, Results and Signficance for LWR Safety," *5th Int. Meeting on Thermal Reactor Safety*, Karlsruhe, September (1984).
184. Denham, M. K. "Experiments of Mixing Molten Uranium Dioxide with Water and Initial Comparisons with the CHYMES Code Calculations," NURETH, Utah (1992).
185. Bird, M. J. "Thermal Interactions Between Molten Uranium Dioxide and Water," *ASME Winter Annual Meeting, Washington* (1981).
186. Ironbridge Museum Pamphlet on the destruction of a furnace in 1801.
187. "The Explosion at Appleby – Frodingham Steelworks – Scunthorpe 4th November 1975,"HMSO (1976).
188. Long, G. et al. "Explosion of Aluminium and Water," Report 2-50-33, Aluminium Research Laboratories, Alcoa (1950).

189. Burgess, D. S. et al. "Hazards of LNG Spillage in Marine Transportation," SRC Report 4105, US Dept. of Interior, Bureau of Mines (1970).

190. Sallack, J. A. "An Investigation of Explosions in the Soda Smelt Operation." *Pulp and Paper* **56** (9), 114–118 (1955).

191. Jennings, A. J. D. "The Physical Chemistry of Safety," *The Chemical Engineer* (Oct 1974).

192. Robinson, C. H. and Fry, C. J. "Steam Explosions Caused by the Contact of Molten Glass and Water," *Proc. Int. Seminar on Chemistry and Process Engineering for High Level Waste Solidification*, KFA Julich (1981).

193. Epstein, L. F. "Metal-Water Reaction: VII Reactor Safety Aspects of Metal-Water Reactions," AEC Report GEAP 3335 (1960).

194. Laker, D. and Lennon, Jr. A. "Explosions of Molten Aluminium and Water," Battelle Memorial Institute Ohio (1970).

195. Cho, D. E. et al. "Some Aspects of Mixing in Large Mass, Energetic Fuel-Coolant Interactions," *Proc. Int. Meeting on Fast Reactor Safety and Related Physics, Chicago* **4**, 1852 (1976).

196. Berthoud, G. "Synthèse des études concernant l'interaction combustible réfrigérant dans les réacteurs à neutron rapides," Note STR/LML/92-93 (1992).

197. Fletcher, D. F. "Large Scale Mixing Simulations Using CHYMES,"AEA Technology – P840 (1988).

198. Corrandini, M. L. "Phenomenological Modelling of the Triggering Phase of Small Scale Steam Explosion Experiments," *Nuclear Science and Engineering* **78**, 154 (1981).

199. El-Genk, M. S. "Molten Fuel-Coolant Interaction Occurring During a Severe Reactivity Initiated Experiment," NUREG/CR-1900 (1981).

200. Papin, J. "The Scarabee Total Blockages Test Series: Synthesis of the Interpretation," *Int. Fast Reactor Meeting*, Snowbird, Utah (1990).

201. McArthur, D. A. "In-Core Fuel Freezing and Plugging Experiments: Preliminary Results of the Sandia TRAN Series 1 Experiments," NUREG/CR-367 (1984).

202. Knowles, J. B. "MFCI and Q^*-events in Fast Reactors," *Canadian Nuclear Soc. Annual Conference*, Saskatoon (1991).

203. Courant, R. and Friedrichs, K. O. "Supersonic Flow and Shock Waves," Interscience, 1956.

204. Leidenfrost, J. G. "On the Fixation of Water in Diverse Fire," Translation of a Latin 1756 discourse in *Int. J. Heat and Mass Transfer* **9**, 1153 (1966).

205. Spivak, V. P. "*Metastable Liquids*", Halsted Press, 1974.

206. Knowles, J. B. "A Mathematical Model of Vapour Film Destabilisation," *AEEW Report* 1933 (1985).

207. Naylor, P. "An Experimental Study of Triggered Film Boiling Destabilisation," *AEEW Report* 1929 (1985).

208. Schmidt, E. "*Properties of Water and Steam in SI Units*", Springer Verlag (1969).

209. Vargaftik, N. B. "*Tables of the Thermophysical Properties of Liquids and Gases – in Normal and Dissociated States*", 2nd Edition, Halstead, 1975.

210. Tabor D. "*Gases, Liquids and Solids – And Other States of Matter*", 3rd Edition, Cambridge, 2000.

211. Chaster, M. "Second Sound in Solids," *Physics Rev.* **131**, 3013 (1963).

212. Inoue, A. et al. "Destabilisation of Film Boiling due to Arrival of a Pressure Shock," *ASME J. of Heat Transfer* **103**, 465 (1981).

213. Rudge, T. "The Status of the FRAX-5B Whole Core Accident Code," *Proc. Int. Fast Reactor Safety Meeting* **2**, 113 (1991) at Utah, USA.

214. Lord Rayleigh, "On Reflexion from Liquid Surfaces in the Neighbourhood of the Polarising Angle," *Phil. Mag.* **3**, 1 (1892).

215. Raman, C. V. and Ramdas, L. A. "On the Thickness of the Optical Transition Layer in Liquid Surfaces," *Phil. Mag.* **33**, 220 (1927).

216. Sharp, D. H. "An Overview of Rayleigh-Taylor Instability," *Physica* **12D**, 3 (1984).

217. Bellman, R. and Pennington, R. H. "Effect of Surface Tension and Viscosity on Taylor Instability," *Quart. Anal. Maths* **12**, 151 (1954).

218. Sparrow, E. M. and Cess, R. D. "*Radiation Heat Transfer*", McGraw, 1978.

219. McAdams, W. H. *Heat Transmission* (Chapter 4 by Hottel, H. C.); McGraw (1954).

220. Bober M. et al. "Spectral Reflectivity and Emissivity Measurements of Solid and Liquid Uranium Dioxide at 458, 514 and 647 nm as a Function of Polarisation and Angle of Incidence," KfK Report No. 3023 (1980).

221. Curico, J. A. "The New Infra-Red Absorption Spectrum Of Liquid Water," *J. Optical Soc. of America* **41**, 302 (1951).

222. Robertson, C. W. and Williams, D. "Lambert Absorption Coefficients in the Infra-Red," *J. Optical Soc. of America* **61**, 1316 (1971).

223. Hale, G. M. et al. "Influence of Temperature on the Spectrum of Water," *J. Optical Soc. of America*, **62**, 1103 (1972).

224. Carslaw, M. S. and Jaeger, J. C. "*Conduction of Heat in Solids*", 2nd Edition, Oxford, 1959.

225. Inoue, A. and Aoki, S. "Study on Transient Heat Transfer of Film Boiling Due to the Arrival of Pressure Shock," *Proc 7th Int. Heat Transfer Conf. Munich*, FB39, 403 (1982).

226. Corradini, M. L. "Modelling Film Boiling Destabilisation Due to a Pressure Shock Arrival," *Nuclear Sc. and Eng.* **84**, 196 (1983).

227. Jordan, E. C. "*Electromagnetic Waves and Radiating Systems*", Constable (1953).

228. Funikawa, O. et al. "Experimental Study of Heat Transfer through Cover Gas in LMFBRs"; *3rd Int. Conf. on Liquid Metal Engineering and Technology in Energy Production*, Oxford (1984) **1**, Paper 92, 451.

229. Rose, J. W. and Nikejad, J. "Interface Matter Transfer: An Experimental Study of the Condensation of Mercury," *Proc. Royal Soc. A*, **378** 305 (1981).

230. Kroger, D. G. and Rohsenow, W. M. "Film Condensation of Saturated Potassium Vapour," *Int. J. of Heat Mass Transfer* **10**, 1891 (1967).

231. Fedorovich, E. D. and Rohsenow, W. M. "The Effect of Vapour Subcooling on Film Condensation of Metals," *Int. J. of Heat Mass Transfer* **12**, 1525 (1968).

232. Mills, A. F. and Seban, R. A. "The Condensation Coefficient of Water," *J. of Heat Transfer* **10**, 1815 (1967).

233. Nabavian, K. and Bromley, L. A. "Condensation Coefficient of Water," *Chem. Eng. Science*, **18**, 651 (1963).

234. Berman, L. D. "Soprotivlenie na granitse, Razdela faz pri plenochnoi Kondensatzii para nizkogo Dableniya", Tr. Vses, N-i Konstrukt in-t Khim Mashinost **36**, 66 (1961).

235. Johnstone, R. K. M and Smith, W. "Rate of Condensation or Evaporation During Short Exposures of a Quiescent Liquid," *Proc. 3rd Heat Transfer Conference* **2**, 348 Chicago (1965).

236. Berthoud, G. et al. "Experimental Collapse of Large Bubbles of Hot Two-Phase Water in Cold Water: The Excobulle Program," *ASME Heat Transfer Conference*, Chicago (1984), Paper 84-HT-18.

237. Walsh, J. M. et al. "Shock Wave Compression of Twenty-Seven Metals. Equations of State for Metals," *Physics Rev.* **108**, 196 (1957).

238. Richtmeyer, R. D. and Morton, K. W. "*Difference Methods for Initial Value Problems*", 2nd Edition, Interscience, 1967.

239. Abramovitz, M. and Stegan, I. (eds) "*Handbook of Mathematical Functions*", 7th Edition, Dover, 1970.

240. Roberts, J. K. "*Heat and Thermodynamics*", Blackie, 1928.

241. Schrage, R. W. "*A Theoretical Study of Interface Mass Transfer*", Columbia University Press, 1953.

242. Sukhatme, S. P. and Rohsenow, W. M. "Heat Transfer During Film Condensation of a Liquid Metal," *ASME Journal of Heat Transfer* **88c**, 19 (1966).

243. Derewnicki, K. P. and Hall, W. B. "Homogeneous Nucleation in Transient Boiling," *Proc. 7th Int. Heat Transfer Conference* **4**, 9 (1982).

244. Board, S. J. et al. "Detonation of Fuel-Coolant Explosions," *Nature* **254**, 319 (1975).

245. Popov, S. G. et al. "Properties of UO_2" Report ORNL/TM-2000/351.

246. Mizuta, H. "Fragmentation of Uranium Dioxide after a Molten Uranium Dioxide–Sodium Interaction," *J. of Nuclear Sc. and Technology* **11**, 480 (1974). Also internal AEEW reports.

247. Doob, J. L. "*Stochastic Processes*", Wiley, 1953.

248. Fry, C. J. and Robinson, C. H. "Experimental Observations of Propagating Thermal Interactions in Metal-Water Systems," *4th CSNI Specialists Meeting on FCI in Nuclear Safety Bournemouth* (1979), Paper FCI/P15, **2**, 329.

249. Royl, P. "PBDOWN: A Computer Code for Simulation of Core Material Discharge and Expansion in the Upper Plenum in an Unprotected Loss of Flow Accident in an LMFBR," KfK Report 01-02-06 PC46C (1985).

250. Breton, J-P. "Some CEA Studies Relating to Core Expansion. The Caravelle Experiments and the IRIS Code"; pp. 317–347 in A.V. Jones (ed.) *Multiphase Processes in LMFBR Analysis*, Harwood, 1984.

251. Reetz, A. "Der Blasencode BERTA Teil 1: Thermodynamik der Reaktionblase," Interatom Report 70-02877 (1984).

252. Reynolds, A. B. and Berthoud, G. "Analysis of Excobulle Two-Phase Expansion Tests," *Nuclear Eng. and Design* **67**, 83 (1981).

253. Reynolds, A. B. et al, "Bubble Behaviour in LMFBR Core Disruptive Accidents," NUREG/CR-2603 (1981).

254. Corradini, M. L. et al. "The Effects of Sodium Entrainment and Heat Transfer with Two-Phase UO_2 during a Hypothetical Core Disruptive Accident," *Nuclear Sc. and Eng.* **73**, 242 (1980).

255. Rose, J. W. Private communications (1990-91).

256. Landau, L. D. and Lifshitz, E. M. "Fluid Mechanics," vol. 6 in *A Course of Theoretical Physics*, Pergamon, 1979.

257. Walsh, J. E. "The MAC Method: A Computing Technique for Solving Viscous, Incompressible Transient Flow Problems Involving Free Surfaces,"LA-3425 (1960).

258. Reed, K. I. "Experimental Investigation of Turbulent Mixing by Rayleigh-Taylor Instability," *Physica* **12D**, 45 (1984).

259. Emmons, H. W. et al., "Taylor Instability of Finite Surface Waves," *J. Fluid Mechanics* **7**, 177 (1960).

260. Lewis, D. J. "The Instability of Liquid Surfaces when accelerated in a direction perpendicular to their Planes," *Proc. Royal Soc. A* **201**, 192 (1950).

261. Cole, R. L. and Tankin, R. S. "Experimental Study of Taylor Instability," *Physics of Fluids* **16**, 1810 (1973).

262. Young, D. L. "Numerical Solution of Turbulent Mixing by Rayleigh-Taylor Instability," *Physica* **12D**, 32 (1984).

263. Jones, A. V. "Numerical Simulation of Liquid Entrainment by the Rayleigh-Taylor Mechanism," *4th Miami Int. Symposium of Multiphase Transport and Particulate Phenomena* (1986).

264. Ploeger, D. W. and Cagliostro, D. J. "Development and Characterisation of a Liquid-Vapour Bubble Source for Modelling HCDA Bubbles," Tech. Report 2, PYU 2939, SRI (1977).

265. Christopher, D. M. "Transient Development of a Two-Phase Jet," Masters Thesis, Purdue University.

266. Mitchell, J. P. Private communication at AEEW (June 1990).

267. Schutz, W. et al. "Wasser-Simulations experimente zum Instantenem Quellterm biem schweren Brutreaktorstörfall," KfK Report No. 4249 (1987).

268. Schlichting, H. *Boundary-Layer Theory*, 6th Edition, McGraw, 1968.

269. Staniforth, M. G. "The Use of SIMMER-II in Fast Reactor Loss of Flow Studies," Int. Workshop KfK (1983).

270. Arnold, L. A. and Knowles, J. B. "Energy Conservation in SIMMER," see 269 above; also available as AEEW Memo 2059 (1983).

271. Procter, J. F. et al. "Response of Enrico Fermi Reactor to TNT-Simulated Nuclear Accidents," NOLTR-62-207 (1964).

272. Drevon, G. A. V. et al, "Comparison of Pressure Loading Produced by Contained Explosions in Water and Sodium," ANL-7120 p720 (1965).

273. Kendall, K. C. and Adnams, D. J. "Experiments to validate Structural Dynamics Codes used in Fast Reactor Safety Assessments," *BNES Conf. on Science and Technology*, London (1986).

274. Kendall K .C. et al. "A SEURBNUK-EURDYN Calculation for a COVA Experiment Representative of a Prototype Fast Reactor Design," *6th Int. Conf. SMiRT*, Paris (1981), Paper E3/4.

275. Neilson, A. J., personal communication.

276. Lancefield, M. J. "Assessment of CDFR primary containment capability under HCDA loading," *Nuclear Eng. and Design* **100**, 221 (March 1987).

277. Potter, R. et al. "Influence of Roof Motion in LMFBR Containment Loading Studies," *Int. Topical Meeting in LMFBR Safety and Related Design and Operational Aspects*, Lyon (1982), Paper TS9B.

278. Wood, S. "Steam Turbine Run-Aways," *Engineer* **207**, 805 (1959).

279. Schueller, G. I. "Impact of Probability Risk Assessment on Containment," *7th SMiRT Conference* (1983), principal lecture for Division J.

280. Duncan, W. J. "*Physical Similarities and Dimensional Analysis*", Arnold, 1952.

281. Clark, L. A. "Crack Similitude in Reinforced Micro-Concrete," in Garas, F. K. and Armer, F. S. T (eds), *Reinforced and Prestressed Microconcrete Models*, Construction Press, 1971, p. 77.

282. Evans, D. J. and Clark, J. L. "A Comparison between the Flexural Behaviour of Small-Scale Micro-concrete Beams and that of Prototype Beams," *Techn. Report Cement and Concrete Association* (March 1981).

283. Mainstone, R. J. "Properties of Materials at High Rates of Straining or Loading," *Materiaux et Constructions* **44(B)**, 102 (1975).

284. Kormeling, H. A. et al. "Experiments on Concrete Under Single and Repeated Uni-axial Impact Tensile Loading," Report 5-80-3, Stevin Laboratory of Delft Technical University (1980).

285. Communicated by Dr. S. Wicks, Head of Impact Technology Department, AEA Technology, Winfrith Technology Centre, Dorset.

286. Neilson, A. J. "An Assessment of the Formulae for Predicting the Damage to Reinforced Concrete Barriers by Flat-Nosed Non-Deforming Missiles," in Neilson, A.J. (ed), "TASD Contributions to the 6th Int. Conf. SMiRT" (1981).

287. Ohte, S. et al. "The Strength of Steel Plates Subjected to Missile Impact," *6th Int. Conf. SMiRT* (1981), Paper 57/10.

288. Canfield, J. A. and Clator, I. G. "Development of a Scaling Law and Techniques to Investigate Penetration in Concrete," Report 2057, Naval Weapons Laboratory (1966).
289. Knowles, J. B. and Neilson, A. J.;unpublished work at AEEW (1982).
290. Irons, B. M. and Razzaque, A. "*Mathematical Aspects of the Finite Element Method with Applications To Partial Differential Equations*", Aziz, A. K. (ed), Academic Press, 1973.
291. Strong, G. and Fix, G. J. *An Analysis of the Finite Element Method*, Prentice-Hall, 1973.
292. Reid, J. K. "Partial Differential Equations," AERE Course Notes (1979).
293. Irons, B. H. "A Frontal Solution Program for Finite Element Analysis," *Int. J. of Numerical Methods in Engineering* **2** (1970).
294. Krutzik, N. "Analysis of Aircraft Impact Problems," in Lecture Notes on "Advanced Structural Dynamics" at JRC Ispra (1978).
295. Drittler, K. and Gruner, P. "Calculation of the Total Force Acting on a Rigid Wall by Projectiles," *Nuclear Eng. and Design* **37**(2), 231 (1976).
296. Broadhouse, J. "Finite Element Analysis," Impact Technology Department, AEA Technology, Winfrith.
297. Macbeth, R. V. "The Burnout Phenomenon in Forced-Convection Boiling" in *Advances in Chemical Engineering* **7**, Academic Press, 1968.
298. "The Safety of Nuclear Power: Strategy for the Future," *Proc. IAEA Conference Vienna* (1991).
299. MacBeth, R. V. "The Effect of Crud Deposits on Frictional Pressure Drops in a Boiling Channel"; AEEW Report 767 (1972).
300. "Light Water Reactor Materials," *ANL*, Feb. 2010.
301. Knowles, J. B. and Fox, P. F. "Local Heat Flux Variations and Burn-out with Heterogeneous Nuclear Reactor Fuel," AEEW Report 748 (1972).
302. Corradini, M. L. "Advanced Nuclear Energy Systems: Heat Transfer Issues and Trends," *Rohsenow Symposium on Future Trends in Heat Transfer* at MIT (2003).
303. Schulenberg, T. "Natural Convection Heat Transfer below downward facing Surfaces," *Int. J. of Heat-Mass Transfer* **28**, 467 (1985).
304. Idel'chik, I. E. "*Handbook of Hydraulic Resistance*", translated for the US Dept. of Commerce; Springfield VA (1966).
305. Nimkar, M. P. "Heat Transfer by Natural Conviction in Two Vertical and One Horizontal Plate," *Int. J. of Engineering and Technology* **3**, No 2, 1008 (2011).
306. Aksan, N. "Selected Examples of Natural Circulation for Small Break LOCA and some Severe Accidents," *IAEA Course on Natural Circulation in Water-Cooled Nuclear Power Plants*, Trieste (2007).
307. Malet, J. et al. "Scaling of Water Spray in Large Enclosures – Application to Nuclear Reactor Spraying Systems," *10th Int. Topical Meeting on Nuclear Reactor Thermal Hydraulics*, South Korea (2003).
308. Lefevre, A. H. "*Atomization and Sprays*", Taylor and Francis, 1989.

309. Zuber, N. et al. "An integrated structure and scaling methodology for Severe Accident technical issue resolution: Development of Methodology," *Nuclear Eng. and Design* **181**, 1 (1998).
310. Rao, D. V. et al. "Knowledge Base for the Effect of Debris on PWR Emergency Core Cooling Sump Performance," LA-UR-03-0880 (2003).
311. Malet, J. et al "Filmwise condensation applied to containment studies: Conclusions of the TOSQAN Air-Steam Condensation Tests," *11th Topical Int. Meeting of Nuclear Reactor Thermal Hydraulics*, Avignon (2005).
312. Miroslav, B. et al "Simulations of TOSQAN containment spray tests with combined Eulerian CFD and droplet tracking model," *Nuclear Eng. and Design* **239**, 708 (2009).
313. Movahed, M. A. and Travis J. R. "Assessment of Gas Flow Spray Model based on the Calculations of the TOSQAN experiments 101 and 113"; *OECD/NEA/IAEA Workshop*, Washington DC (2010).
314. Moore, R. V. "The Dounreay Prototype Fast Reactor," Nuclear Eng. International, (August 1971).
315. Sohal, M. S. and Siefken, L. J. "A Heat Transfer Model for a Stratified Corium-Metal Pool in the Lower Plenum of a Nuclear Reactor"; Idaho National Laboratory Report INEEL/EX-99-00763 (1999).
316. Akers, D. W. et al., "Examination of Relocated Fuel Debris Adjacent to the Lower Head of the TMI-2 Reactor Vessel," NUREG/CR-6195 (1994).
317. Hofmann, P. et al., "Reactor Core Materials Interactions at Very High Temperatures," *Nuclear Technology* **87** (August 1989).
318. Kelkar, K. M. et al. "Computational Modelling of Turbulent Natural Convection in Flows Simulating Reactor Core Melt," Report to Sandia National Laboratory from Innovative Research, Inc. (1993)
319. Siefken L. J. et al., "SCDAP/RELAP5-3D User Manual"; 5 Volumes, Idaho Nat. Eng. and Environmental Laboratory Report 00589 (2002).
320. "Radioactive Waste Management," World Nuclear Association Briefing Papers (April 2012).
321. "Natural Fission Reactors – The Oklo Phenomena," *UKAEA Atom* **391**, 30 (1989).
322. BBC News, January 26, 2013.
323. BP plc Annual Report (2012) and www.bp.com/unconventionalgas
324. Helm D. "Natural Capital," J Roy Soc for Arts, Commerce and Production Spring 2013.

Index

Nuclear Electric Power: Safety, Operation, and Control Aspects, First Edition.
J. Brian Knowles.

© 2014 John Wiley & Sons, Inc. Published 2014 by John Wiley & Sons, Inc.

Printed and bound by CPI Group (UK) Ltd, Croydon, CR0 4YY

16/04/2025

14658344-0003